Industry 4.0 Topical Conference 2019

Topical Conference at the 2019 AIChE Spring Meeting
and 15th Global Congress on Process Safety

New Orleans, Louisiana, USA
31 March - 4 April 2019

ISBN: 978-1-5108-8681-0

Printed from e-media with permission by:

Curran Associates, Inc.
57 Morehouse Lane
Red Hook, NY 12571

Some format issues inherent in the e-media version may also appear in this print version.

Printed by Curran Associates, Inc. (2019)

For permission requests, please contact AIChE
at the address below.

AIChE
120 Wall Street, FL 23
New York, NY 10005-4020

Phone: (800) 242-4363
Fax: (203) 775-5177

www.aiche.org

Additional copies of this publication are available from:

Curran Associates, Inc.
57 Morehouse Lane
Red Hook, NY 12571 USA
Phone: 845-758-0400
Fax: 845-758-2633
Email: curran@proceedings.com
Web: www.proceedings.com

TABLE OF CONTENTS

554288 Intelligent Processes: Leveraging the Data Revolution for Decision and Control

We are currently at the dawn of what is considered the fourth industrial revolution. This revolution is driven by a combination of internet connected devices, large volumes of data, significant computing power and major algorithmic advances in artificial intelligence. It presents a unique opportunity for process industries to interpret data in new ways and derive information necessary to enable intelligent decisions at various levels of operation.

We will start with an introduction to major recent breakthroughs in artificial intelligence and describe how they can be used to solve problems in process modeling, fault detection and diagnosis and control. In particular, we will take the audience through a journey of process data analytics that includes traditional multivariate statistical approaches as well as more modern tools such as variational auto encoders and deep neural networks. This will be followed by a quick tour of reinforcement learning and its potential to change the decision and control paradigm as we know it today. The talk will be interspersed with several relevant industrial and simulated examples.

The Digital Process Unit

Refinery Operations Monitoring and Optimization Using Real-Time Data Collection and Advanced Models

Nevin Whalley, David de Gruyl, Matthew Hutchinson - Axens North America

INTRODUCTION

In the oil and gas industry, refiners continue to seek incremental improvements to existing assets in order to remain competitive. High speed data networks, cloud-based systems, and advanced data compilation tools are leading the way of cost effective techniques used in the industry to monitor and optimize its operating assets. Axens combines these solutions with real-time, advanced modeling in order to provide the greatest impact to process safety, operational excellence, and operating costs. The Axens' digital suite includes:

- Connect'In™: a remote unit monitoring system displaying process data analyzed through high-fidelity kinetic models for engineers, operators, and management on an online platform.
- Advanced Process Control: a computational logic for multivariable control of unit performance based kinetic and thermodynamic models
- Real-Time Optimization: a next-generation, model-based solver for improved refinery performance based kinetic and thermodynamic models

This paper will present case studies, demonstrations, and economics highlighting the value added by integrating advanced real-time unit modelling into key performance indicator matrices, advanced process control, and optimized performance.

CONNECT'IN™

Axens' Connect'In™ is a remote unit monitoring platform that utilizes immediate process data to deliver real-time, custom made performance indicators using high fidelity models. This is accomplished by connecting a direct data transfer link, making the results easily and securely accessible through an online tool, and applying high fidelity kinetic models.

Automatic transfer and direct access of unit data allows Axens to provide faster and more proactive technical assistance for unit engineers to avoid sub-optimal operation and economic losses due to lost time. Tailor-made key performance indicators (KPIs) such as process normalizations are regularly updated and linked with alerts to notify the plant of any immediate need for unit optimization. Additionally, predictive kinetic models are utilized to evaluate future catalytic performance and estimate end of life cycles based on changes in unit feed quality or operation severity. Connect'In™ unit monitoring also expands beyond catalytic support through the integration of a process heater monitoring tool supported by the models and expertise from Axens' subsidiary, Heurtey-Petrochem.

The next generation of process unit monitoring will require teams comprised of unit engineers, catalyst supplier technical service engineers, and corporate subject matter experts, all accessing the same database and optimizing unit performance proactively rather the reactively. The Connect'In™ platform provides a consistent and automated launchpad for collaboration and discussion on improving unit operation.

Case Studies Justifying Connect'In™ Deployment

A semi-regen reformer was displaying decreased performance as it reached the end of run reactor inlet temperature. However, even after turnaround, the same poor performances were recorded and Axens was requested to investigate the issue and find the root cause. Axens technical support noticed that the trend of the cyclohexane to methylcyclopentane ratio in the product had risen sharply months before the unit shutdown, indicating that a leak in the reactor feed/effluent heat exchanger was the source of the poor performance. The Connect'In™ platform utilizes real-time data calculations and trend comparisons to alert both the unit and Axens engineer of such issues as they happen, compared to troubleshooting well after the fact. Without Connect'In™, the refiner for this case suffered economic losses in the range of $2-3 million per year.

A CCR unit was experiencing a loss in performance without any major operational changes. Based on the comparison of unit performance to Axens' kinetic models, the

Axens technical support engineer noted a decline in unit yields and possible catalyst deactivation. After the alert, Axens worked closely with the site engineers to determine that Silicon break-through from the NHT was deactivating CCR performance and occurring as a result of a recent shift in crude slate. Connect'In™ brings Axens' high fidelity models directly to the client in order to monitor catalyst performance in real-time, allowing for faster response times to preserve activity through the catalyst's life. These same kinetic models can also be used to predict how variations to the unit, such as feed quality, will affect the catalyst cycle length before such changes are implemented.

Another case study highlighting the benefits of proactive data transfer and analysis comes from a Prime-G+™ FCC gasoline desulfurization unit which experienced an unexpected 2 point octane loss. The original support strategy consisted of Axens Technical Services requesting detailed process data and lab samples to perform a troubleshooting study to determine the root cause for the octane loss. For this case, the amount of delayed time for the data request, transfer, and analysis cost the refinery nearly $1 million. This economic loss could have been avoided with the early detection and direct data link benefit of Connect'In™. This absolute connectivity makes Axens engineers available to provide proactive technical assistance and troubleshooting from any location.

Data Visualization

The primary technology platform is the Connect'In™ Dashboard, a summary of the entire unit through several customizable performance indicators highlighting any immediate threats to product quality. Axens collaborates with clients to create the list of high priority graphs to include in the Dashboard that outline overall unit trends. Each graph on the Dashboard includes customized alerts that compare the last data point on the graph or an average of the trend with an Axens recommended operating range to advise the client on operational changes to optimize the unit performance.

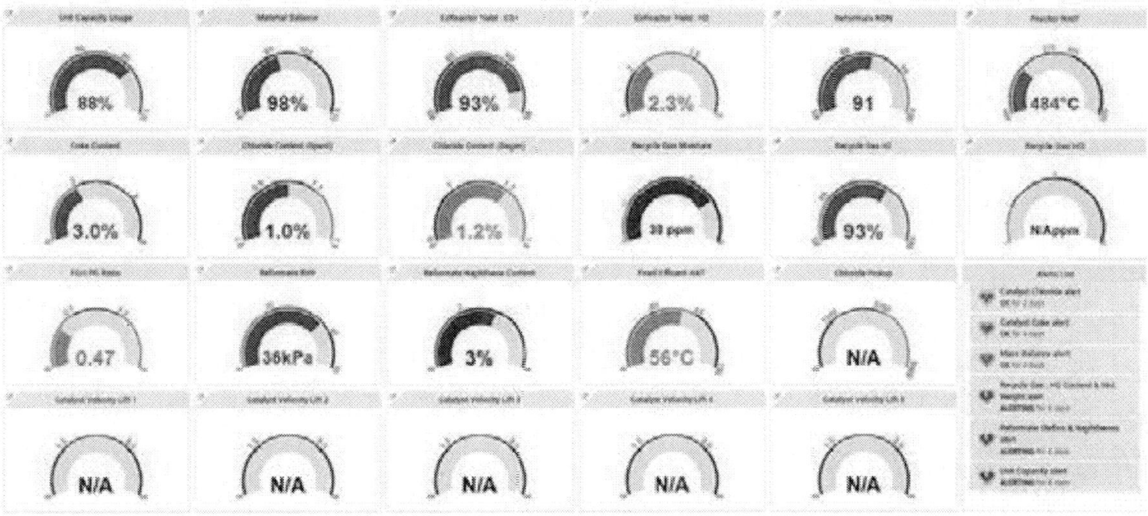

Figure 1: Connect'In™ Dashboard for a summarized look at unit performance

Figure 2: Connect'In™ Dashboard of normalized data with performance alerts based on process data and trends

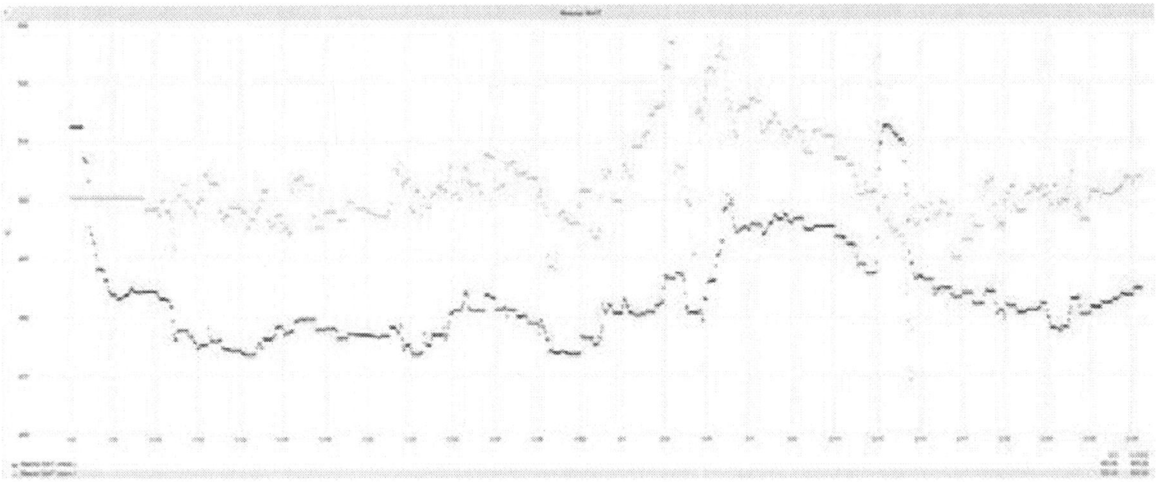

Figure 3: Detailed graph with individual data points, adjustable axis, and selective data legend

Besides accessing the detailed graphs through the Dashboard, users can access a multitude of different plots through pulldown menus organized by unit sections or operating category. Within these categories, users will be able to observe the Key Performance Indicator (KPI) trends calculated using Axens kinetic and predictive models.

Axens continues to provide data monitoring flexibility by voluntarily passing calculated results (normalized data and kinetic model results) back into the client's historical database. If implemented, this feature allows the user to display the results of the model predictions alongside the real-time operating data on any in-house system to maximize the value of their own data combined with Axens' calculated results.

Fast and Secure Data Transfer

Unit monitoring requires large quantities of process data to be filtered and analyzed as quickly as possible. Cloud-based data collection improves the current unit monitoring system by directly modifying and sharing all the process data from a unit directly to those who need to be aware of its operation in real time. Connecting to this high speed data network drastically decreases lost time between the initial transfer and the final visualization of process data.

Cybersecurity is an important topic to cover when discussing cloud-based technologies and how to protect valuable IP. In response, Axens remains flexible on the method of data capture and reconciliation by using either a turn-key infrastructure solution or an interoperable platform similar to a client's infrastructure. This ensures a secure data transfer system while still complying with the various IT restrictions seen across the industry.

Industrial Deployment

Connect'In™ has been deployed to 9 units across five refineries within the first two years of client implementation. The integration of Connect'In™ into multiple units within the same refinery has improved the technical support from Axens by centralizing the clients' process data and facilitating discussions with unit engineers across the plant. Another method of implementation, performed by a major US refiner, installs Connect'In™ for the same technology across several refineries in order to perform benchmarking analysis and performance reviews.

ADVANCED PROCESS CONTROL

As product specifications become increasingly stringent, Advanced Process Control (APC) is a low cost and continuously operating solution to maximize profitability.

APC helps clients achieve better and safer plant performance by:

- Extending catalyst performance and service life
- Reducing utility and logistic costs
- Saving energy and decreasing carbon dioxide emissions
- Avoiding the problems caused by off-specification products

While many refiners use matrix based Advanced Process Control, this solution is typically only effective when tuned near the intended operating point. In the event of a change in feed or product specifications, the matrix may require re-tuning and may not function well. If a kinetic model is integrated into the solution, the number of tuning parameters can be drastically reduced, and the system can self-correct for changes in feed properties or product specifications.

Model based inferences are key for the balanced control of economically or environmentally critical final product values. Specifically, by readjusting the important feed properties, a simulation internal to the controller can provide adjustment vectors for the controllers.

A case study from a refinery shows the economic benefit of using APC for a gasoline desulfurization unit. The client was having difficulties maintaining a constant product specification of 20 wppm sulfur, often averaging 17 wppm. This over-treatment was leading to negative performance effects including increased octane loss with high hydrogen and utilities consumption. Axens provided a unique APC solution for the refiner which combined extensive process knowledge, intensive operational expertise, and proprietary models. The APC project was started seven months after the unit was commissioned and took only four months to install. A post-audit was performed six months after the APC commissioning and the economic results were overwhelming.

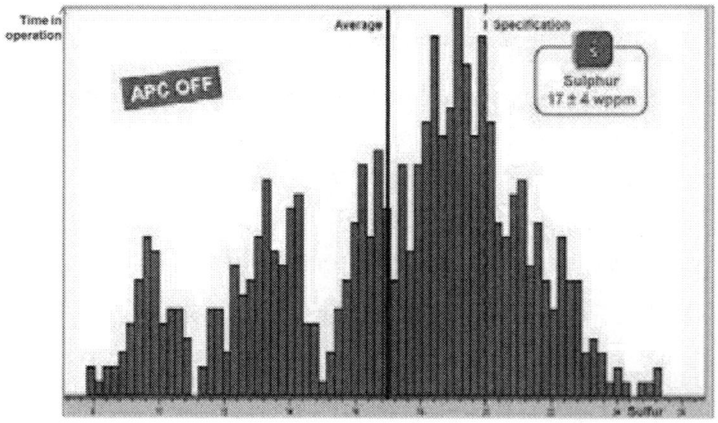

Figure 10: Graphical representations of the product sulfur specifications with the APC off

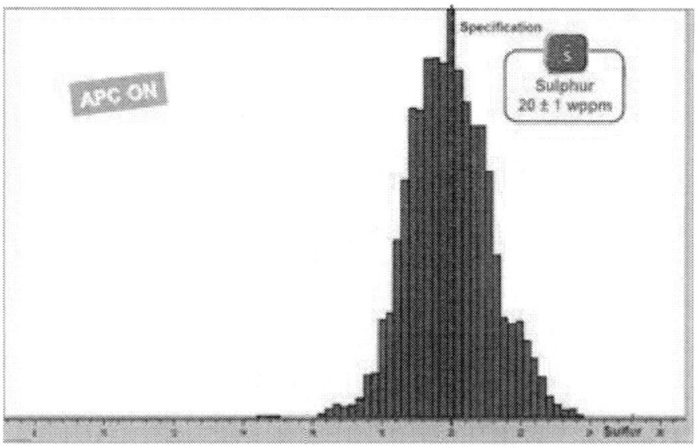

Figure 11: Graphical representations of the product sulfur specifications with the APC on

While using the APC solution, the refiner saw an average product sulfur of 20 wppm with an improvement in octane retention, hydrogen consumption, utilities consumption, and even catalyst cycle length. All of which increase the profitability of the unit by an order of 5-6M$ per year with the cost of the APC being returned within the first year.

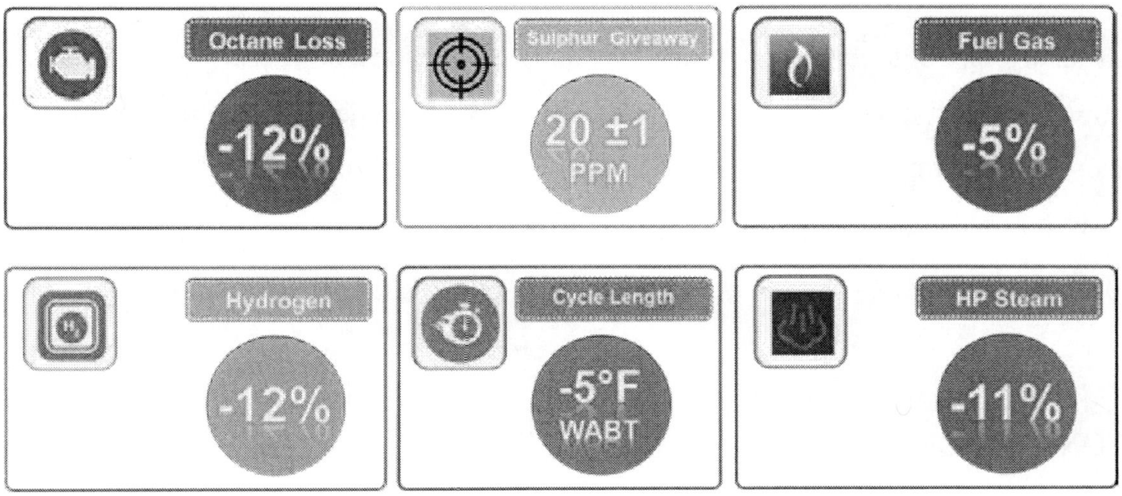

Figure 12: Benefits detailed from post-audit of refinery using APC solution for improved average product sulfur specification

Axens has the ability to manage all phases of APC projects with the use of powerful computational resources such as inferential modules, proprietary software packages, multivariable predictive controllers, and an extensive database of chemical reaction kinetics and thermodynamic properties.

REAL-TIME OPTIMIZATION

While real-time optimization software has been on the market for some time, the integration of rigorous models for reaction kinetics and specific process behaviors allow for the creation of a true digital twin for the purpose of optimizing unit operations, and eventually optimizing the plant-wide and enterprise-wide operations targets.

Rigorous models are developed based on both research and industrial data, and allow for the accurate and realistic modeling of the operation of units, allowing for the improvement of unit and plant profitability.

The integration of model correction based on feedback from actual process data is a leap forward in optimization design, creating an opportunity to implement the most useful models for the optimizer platform.

LP MODEL INTEGRATION

By using the rigorous model-based operation projections built previously, a distilled version of this model can be integrated into the LP modeling system for plant-wide economic optimization. This model provides vector information to the LP model for various operating conditions and target, creating a more accurate view of the operation at these conditions.

CONCLUSION

In the current environment of the oil and gas industry, digital solutions are crucial for refiners to achieve efficient asset monitoring and effective unit optimization. Axens provides a range of customized software solutions combined with its proprietary high fidelity models to improve unit safety and operational excellence. Cloud-based data monitoring through Connect'In™ allows for remote access to process data and calculated KPIs in real-time, improving troubleshooting efficiency and avoiding economical losses. Dynamic operator training simulators educate operators and engineers through interactive lesson-based exercises improving the retention of important knowledge out in the field, emergency scenario response times, and safe normal operation. Advanced process control provides automated multivariable unit operation increasing profitability and reducing loses from poor product quality and utilities waste. The Axens digital suite offers complete and cost-effective solutions to any refinery aiming to proactively improve the operation and performances of current assets.

549847 Next Generation Process Modelling Utilizing Digital Twins

Making timely and effective process scheduling and operational decisions, within an environment of constant change, is a key success factor in today's dynamic market. With recent technology advances, legacy simulation modelling tools are being replaced with state-of-the-art process simulation suites, where one model and one platform services multiple purposes. Not only faster and more reliable, the next generation refining and petrochemical modelling platform can solve complex problems, without the need for a "what-if" trial and error approach.

A next generation process modelling platform is designed to be generally applicable to any modelling need, without requiring extremely specialized skills to develop and maintain. An open platform useable by anyone to express their domain knowledge without specialized programming skills. A platform where the direct application of rigorous mathematical optimization capability can help calibrate a digital twin model and be used to explore a rigorous model decision space.

This digital twin model on an optimization platform provides a simultaneous solution technique to reduce maintenance costs, increase plant utilization, maximize plant performance via energy reduction, quality giveaway minimization and/or yield improvements, and create fast and accurate analytical derivatives for integration with external optimization, planning and scheduling applications to enhance the business portfolio, capturing opportunities in the global markets. All of this is accomplished with a computationally fast “warm start” capability to track and compare model and plant operations in real time. It has been predicted that by 2021, half of large industrial companies will use digital twins, resulting in those organizations gaining a 10% improvement in effectiveness.[1]

This paper presents the use of UniSim Design for real time refining and petrochemical modelling via Digital Twin applications, which supports a faster decision-making process for plant optimization and reliability improvement programs.

[1] https://www.gartner.com/smarterwithgartner/prepare-for-the-impact-of-digital-twins/

547349 Data Characterization, Visualization and Analytics for a Smart Manufacturing Testbed

The emergence of the industrial Internet of Things (IoT) and ever advancing computing and communication technologies have fueled a new industrial revolution. This revolution is happening worldwide to make current manufacturing systems smarter, safer, and more efficient. Although many general frameworks have been proposed for IoT enabled systems and their potential in industrial applications, there is limited literature on demonstrations or testbeds of such systems. Simulation is a powerful tool but the fidelity of the simulated system is limited by the understanding on the system; that is, the model that describes the system. In this work, with industrial IoT still in its infancy, there is not sufficient understanding on the property, capacity and performance of IoT sensors to enable accurate simulation. In addition, there is a lack of systematic study on the characterization and representation of data generated by industrial IoT sensors and data analytics challenges associated with industrial IoT sensor data. These are the major motivations for us to build a smart manufacturing testbed equipped with IoT vibration sensors, and to investigate the capabilities of these IoT vibration sensors for process monitoring by building soft sensors to predict key process variables. We will discuss the details of the testbed setup; investigate the characteristics of the data collected from the IoT vibration sensors; and explore data representation and visualization. The results from soft sensor development and validation will be presented, and summaries and future directions will be provided. We expect that the findings from this work can be generalized and extended to actual manufacturing systems.

551035 Value of Increasing Sample Size for Assessing Defect Concentration

Purity and cleanliness are important quality attributes in many products. Foreign contamination or other defects within a product has a detrimental effort on the real performance of the material in end-use applications. Visual evidence of any defects conveys a negative impression to customers with regard to the product quality. Therefore, in many manufacturing or processing facilities, the product is assessed by an on-line inspection system to quantify the concentration of defects for product classification and release. In most cases, the production rate of the facility greatly exceeds the throughput of the available inspection devices so a representative portion of the material produced is inspected. Inspection devices are available with different achievable inspection rates. Optimizing the value of inspection process requires balancing the increased cost of larger inspection rates with the value gained from improved confidence of the results. The concentration of defects is typically on the order of parts per million, and measurements of low count data have inherently large variability. This results in a meaningful amount of Type I and Type II error (incorrect prime/off-grade decision). The impact of increasing the measured throughput was explored using statistics appropriate for discrete data, and the resulting change in Type I/II error determined to assess the value of increasing the inspection rate.

550287 Disturbance Tracking in Industrial Milk Powder Process Systems

Milk powder quality has become increasingly important in the competitive commodity market, leading to a focus on technologies and systems that can enable the monitoring of quality in real time. A key aspect effecting milk powder quality are is process disturbances, however the effect on milk powder quality is not known. One known disturbance cause is the periodic switching in, and out, of equipment for cleaning during continuous milk powder manufacture, to maintain process hygiene. It was thought that these line swaps were the main contributor to non-steady state operation of the plant. However, although the timing of these line-swaps was known, their magnitude, duration and relative proportion in relation to total disturbance and manufacturing time, was not quantified. Therefore to date it has not been possible to infer whether they could affect quality without further a quantitative data analysis.

In this project, we used both conventional statistical monitoring tools such as statistical control charts, (SPC), and advanced modern regression techniques such as big data analytics including machine learning to analyse historical milk powder process plant data. Both methods performed equally well for tracking the historical process disturbances. The results showed that most disturbances in the total solids were correlated with line swaps, although not all. Furthermore the magnitude of the disturbances experienced during a line swap varied markedly, indicating that there is scope for minimising potential disturbances through specific operational intervention. However, there were a number of short, and long duration disturbances in the total solids that were not linked to line-swaps, indicating other significant mechanisms at play. This analysis helped us to reveal the line swap impacts on the quality variables, e.g. total solids in the concentrated milk; to identify different line swap effects on down-stream units; and to classify the different disturbances identified. This work has also indicated a more thorough analysis if the clean-in-place is warranted in order to hold the total solids content steady.

550591 Some Challenges in Analyzing Manufacturing Data for Process Optimization

In recent years, manufacturers collect and store more data than was previously feasible due to the drop in the costs of both sensor technologies and data storage. With the significant advance of analytics techniques, valuable information could be extracted by analyzing manufacturing data to provide insight into process optimization. This presentation discusses several challenges in data pre-processing and statistical modeling.

With hundreds or thousands of process variables showing different patterns, there are risks in developing scripts to automate the data cleaning process. In addition, switches between products and process maintenance cause difficulties in performing time series analysis to investigate cyclical patterns in the process. Furthermore, the partial least squares (PLS) method can be used to identify relatively important variables for root cause investigation when there are high correlations between process variables. However, with hundreds of process variables being important in the PLS model, extra effort is required to judge the impact of these process variables.

545276 Designing Industry 4.0 Projects for Process Safety

Designing Industry 4.0 projects for process safety requires understanding the wide spectrum of cyber/physical technologies. This includes Advanced Process Control, Artificial Intelligence, Neural Networks, Digital Twins, Machine Learning, Robots /Cobots, and Operational Guidance Systems. Analytics is a major part of this, but which techniques are most appropriate and how do you best ensure success? Developing a strategy to integrate these technologies involves a process which incorporates goals and a methodology for implementation.

Industry 4.0 is well suited for process safety, and there are key benefits for determining the order of deployment. This may vary from company to company depending on goals and based on your company's maturity/capabilities. Finally, how do you best scale analytic deployments across the enterprise? This talk will focus on these issues and the factors to consider when developing an overall strategy.

Jim Petrusich is the co-author of The Industry 4.0 for Process Safety Handbook *(October 2017).* Northwest Analytics is the winner of the 2018 *Frost and Sullivan* Industry 4.0 Award.

550630 Using Industry 4.0 to Create a Pathway to Operational Excellence

I think we all recognize there is journey to get to Industry 4.0 or Refining 4.0. How do we get to the smart refinery? The smart refinery includes the Internet of Things, Cloud computing, cyber-physical process, and Analytics. Teradata is working with companies on their journey to smart refineries by helping companies with their analytics through IoT and cloud computing at scale of machine learning algorithms. Teradata is providing companies with access to the power of analytics by combining their sensor data and ERP data at granular detail level to gain insights on how to improve costs and increase gross margin in their reliability and optimization processes. We will discuss the methods we apply to to bridge the OT (Operational Technology) and IT(Information Technology) worlds together to have impacts on maintenance programs, energy optimization, and operations. We will also discuss how data integration around applications, data silos , ERP systems have to helping companies not only gain insights but also have financial impact to the bottom line at scale for multiple refineries.

554383 *Accelerating Product Innovation at Dow through Multivariate Modeling* (Poster)

In the era of Industry 4.0, manufacturers such as Dow are focused on gaining more value from their historical data than ever before. ProSensus has distinguished itself as a trusted, global leader in Big Data analytics by helping many Fortune 500 companies across numerous industries use their data to innovate for the future.

ProSensus has recently worked with Dow to accelerate innovation on one of their key product lines – silicone antifoams. For product development applications such as this, ProSensus combines powerful multi-block latent variable modeling with constrained optimization. This approach allows clients to simultaneously optimize the selection of raw materials, recipe formulations, manufacturing conditions and costs to reach targeted product performance properties while adhering to custom constraints. A model-based approach allows Dow to rapidly develop and scale up custom silicone antifoam formulations for the dynamic technical and regulatory needs of customers in the pulp and paper, food and beverage, wastewater treatment, metal working, and other industries.

This presentation will examine the approach taken (data assembly, multivariate modeling, model validation, and optimization), challenges encountered, and results obtained. Perspectives from both ProSensus and Dow will be included.

554363 Reinforcement Learning (RL) Based Data-Driven Decision Support System (Poster)

Big data has transformed from a buzzword to a real value creator. In recent years there has been a rising interest in data analytics in the energy industry. The industry and academia have made significant progress in big data analytics. However, there is still a need to use the plant data to gain insights and address the issues related to abnormal situation management. An abnormal situation is defined as "a disturbance or series of disturbances in a process that causes plant operations to deviate from their normal operating state." The disturbance can be a minor near-miss to a catastrophic incident which may cause production losses, environmental damage, injuries or sometimes fatalities if not managed properly. During an abnormal situation the operator is expected to take appropriate action to bring back the process to a normal state. Diagnose and mitigation actions depend on the operator's experience, process complexity and stress levels of the operator.

This paper outlines a systematic framework for application of Reinforcement Learning (RL) during process operations and abnormal situation management. A novel method using Reinforcement Learning is established to address decision making challenges and advise operators during abnormal situations. The system is designed on the expert knowledge from operators and engineers, the knowledge available from previous operation experiences and process plant information. The application of the proposed method can be expanded by connecting the system to Operator Training Simulators (OTS) and train operators for abnormal situation management.

A sequence to sequence time series forecasting model for batch processes with seasonality

Shu Xu, Mark Nixon

Emerson Automation Solutions

1. Introduction

Modern chemical manufacturing processes generally fall into three categories: continuous, batch and semi-batch. Unlike its continuous counterpart where the raw materials are constantly fed into the system and processed into an intermediate or final product continuously, a batch process receives and processes the feedstock by a series of discrete steps. Such a manufacturing process has been widely used in semiconductor, pharmaceutical, specialty chemical and other industries. One example is the fermentation process used in the brewing industry. Because the whole process, consisting of many operating steps, can hardly be operated in a steady state, it poses tough challenges for both equipment and operators to assure product quality over many batch runs. One specific challenge is related to seasonality which stems from equipment aging[1], periodic maintenance[2], catalyst decay [3] and so on. Typical examples of such a behavior involve activity coefficients of reactor catalysts and fouling factors in heat exchangers[4]. Process engineers are challenged to forecast such a seasonality lurking in the system so that they can become proactive in identifying and preventing potential problems.

In this paper, we built a sequence-to-sequence model to forecast the seasonality simulated in the benchmark Penicillin bio-reactor process. The results obtained by such a sequence-to-sequence model are compared with those generated from a linear partial least square (PLS) method and a nonlinear support vector machine (SVM) regression method.

The remainder of this paper is organized as follows. Section 2 briefly introduces methodology information regarding time series forecasting, long short-memory (LSTMs) networks, and an associated sequence-to-sequence model. Section 3 provides case study details including the benchmark Penicillin bio-reactor simulator and simulation conditions. Section 4 summarizes results obtained in the case study. We conclude the paper and discuss future works in Section 5.

2. Methods

2.1. Time series forecasting

During batch operations, abundant informative data can be collected in the form of time series samples with regular sampling periods [5]. As a result, forecasting methods in time series analysis, which take advantage of dependence between observations in the time dimension to predict the future, can be applied in process seasonality forecasting [6]. For example, classical methods include autoregression, autoregressive integrated moving average (ARIMA), simple exponential smoothing (SES), etc. In addition, time series forecasting can also be reframed as a supervised learning problem. By reframing the time series data via a lag method, various linear and nonlinear supervised learning algorithms can be used, including linear regression, support vector machine, neural networks, or a sequence-to-sequence model with long short-term memory (LSTM) networks as discussed in this paper.

2.2. LSTM network

Long short-term memory (LSTM) networks belong to the category of recurrent neural networks (RNNs, see Figure 1) where loops and chains exist to allow information to persist. In Figure 1, an input x_t is fed into a regular neural network block A which produces an output h_t. An RNN distinguishes itself from its regular counterpart in that each neural network unit in the RNN can pass information either to itself in a loop or to a successor to form a chain. Although a normal RNN can track the short-term dependence between observations in the time series, it struggles to learn the "long-term" connections [7]. To overcome such a challenge, the LSTM network is proposed[8] (see Figure 2).

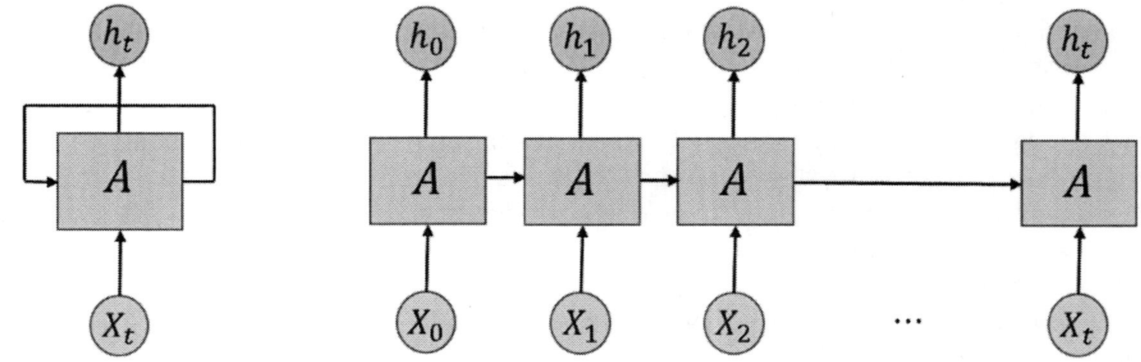

(a) An RNN with loop (b) An unrolled RNN

Figure 1 Recurrent neural network demonstration

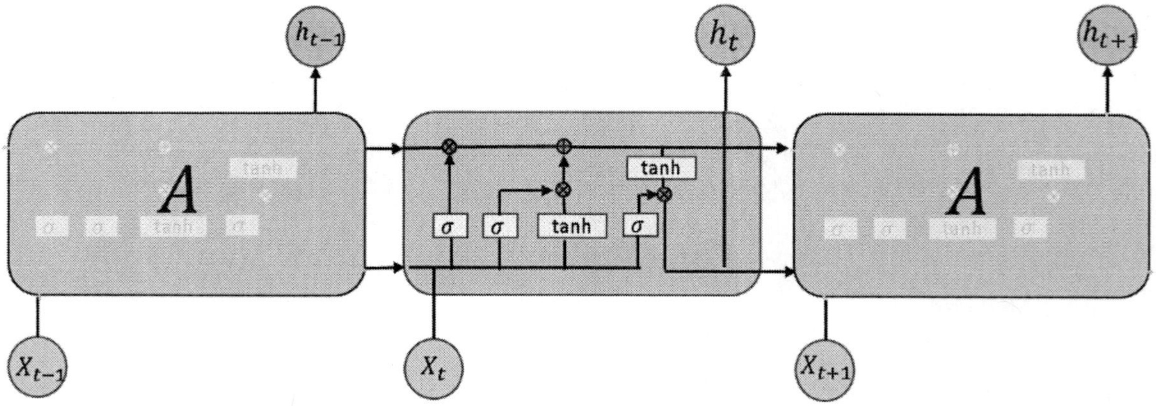

Figure 2 LSTM network demonstration

Although the general form of the LSTM network is akin to that of a regular RNN, each cell contains four neural network layers that interact in a way that enables them to remember information for long periods of time. A detailed description of the structure of each cell follow.

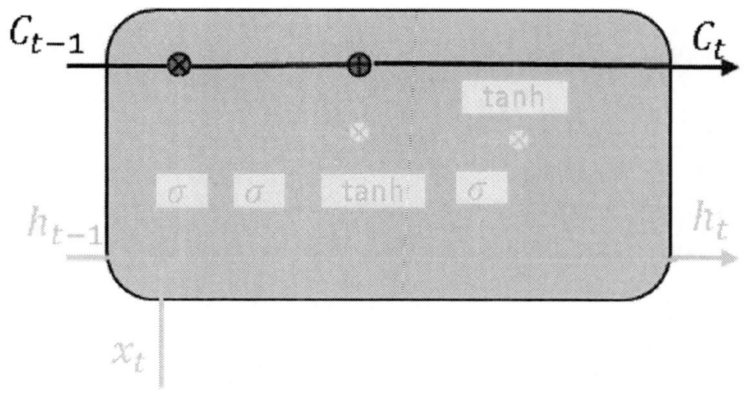

Figure 3 Cell state in an LSTM cell

As illustrated in Figure 3, a horizontal line running across the top of the cell with only linear pointwise operation indicates a potentially free flow of information without any cell status changes. In addition, the amount of cell status information gain or loss is controlled by the following gates in Figure 4, which are comprised of a sigmoid neural net layer and a pointwise multiplication operation. Because the sigmoid layer outputs a number between 0 and 1, it regulates how much information can come through. While a value of 0 indicates closing the gate, a value of 1 means letting the information flow through freely.

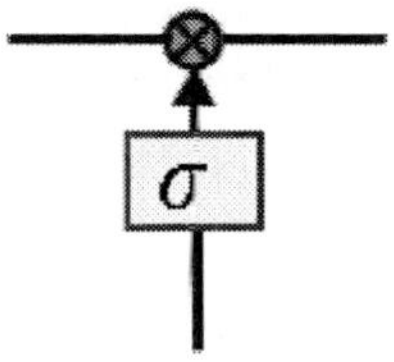

Figure 4 Gate in an LSTM cell

Figures 5 through 8 provide a step-by-step walk-through of the four-layer structure inside each LSTM cell. As shown in Figure 5, the first layer, named the "forget gate layer", decides whether to keep the previous cell state C_{t-1} or not. For example, if the forgetting factor f_t calculated by gate Equation (1) equals 0, the previous cell state is completed discarded (or "forgotten") after an element-wise multiplication.

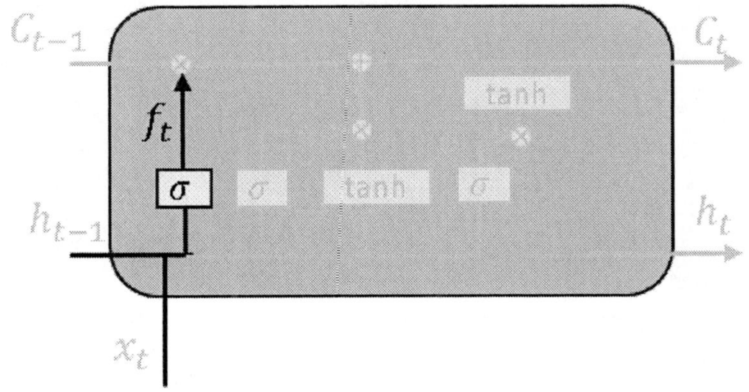

Figure 5 Forget gate layer in an LSTM cell

$$f_t = \sigma\left(W_f \cdot [h_{t-1}, x_t] + b_f\right) \tag{1}$$

The next state update layer consists of two sub-layers. Out of these layers, the "input gate layer" decides whether we need to update the cell state with the new value produced by the "new value layer", as shown in Figure 6. The associated gate and value equations are listed in Equations (2) and (3) respectively.

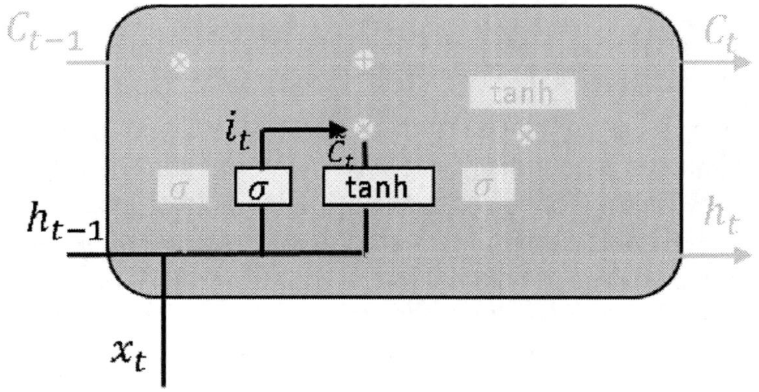

Figure 6 State update layer in an LSTM cell

$$i_t = \sigma(W_i \cdot [h_{t-1}, x_t] + b_i) \tag{2}$$

$$\tilde{C}_t = tanh(W_C \cdot [h_{t-1}, x_t] + b_C) \tag{3}$$

Now since we have both the previous cell state C_{t-1} and the new state update value $\tilde{C}_t$, the new cell state C_t can be calculated by Equation (4). Figure 7 shows the state update.

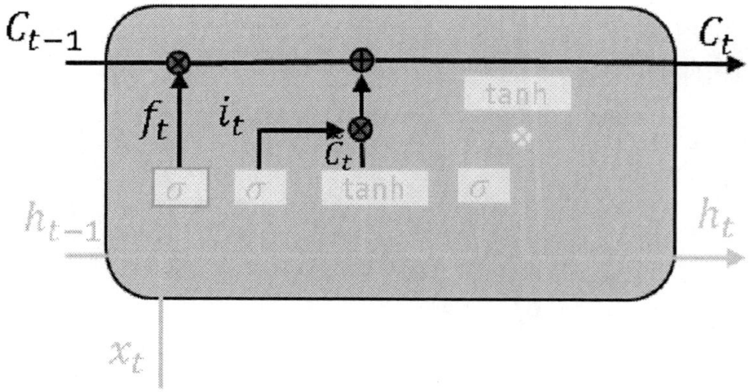

Figure 7 State update operation in an LSTM cell

As shown in Equation (4), the new state C_t equals the sum of an old state C_{t-1} multiplied by the forgetting factor f_t, and a new state update $\tilde{C}_t$ scaled by a factor i_t.

$$C_t = f_t * C_{t-1} + i_t * \tilde{C}_t \quad (4)$$

Finally, the output of the cell is decided by the output layer shown in Figure 8, including a sigmoid gate which decides whether we need to deliver a scaled version of the cell state C_t by "tanh" function, i.e., the h_t. Associated equations are shown in Equations (5) and (6).

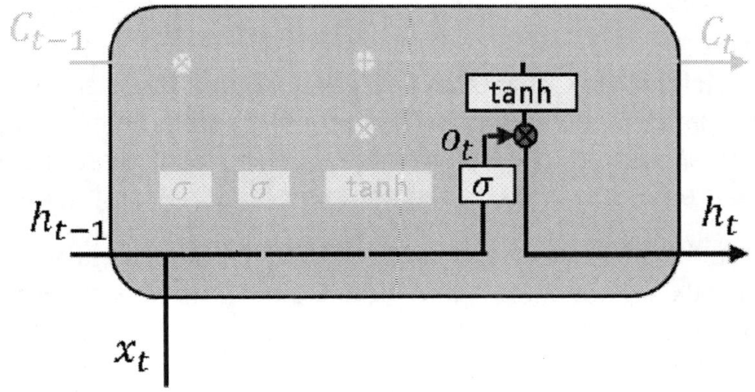

Figure 8 Output layer in an LSTM cell

$$o_t = \sigma(W_o \cdot [h_{t-1}, x_t] + b_o) \quad (5)$$

$$h_t = o_t * tanh(C_t) \quad (6)$$

In our study, a sequence to sequence (seq2seq) model with LSTM cells is used for time series forecasting. The model includes an encoder converting an input sequence X_t to corresponding state vectors which contain transformed context for each input point x_t, and a decoder which takes context vectors and internal hidden states as inputs to generate an output sequence Y_t.

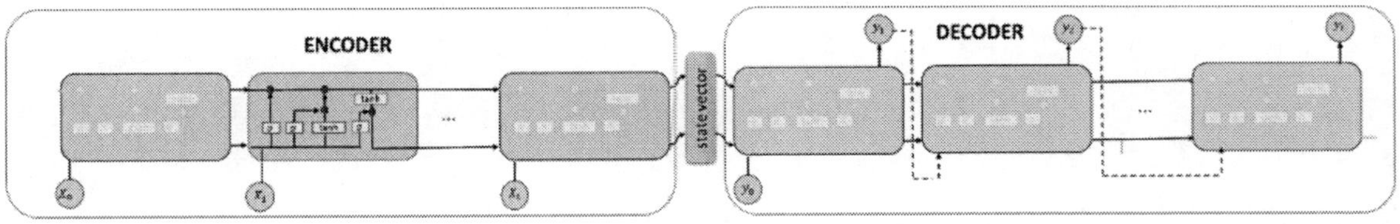

Figure 9 A seq2seq model with LSTM cells

3. Case study

3.1. The Benchmark Penicillin Bio-reactor simulator

A lumped-parameter dynamic batch-fed Penicillin model developed by Birol et al. [9] is used as the basis for the current study. The original MATLAB code for the simulation program can be downloaded from Dr. Cinar's website (www.chee.iit.edu/~cinar/software.htm), and modifications have been made on the key kinetic parameter to simulate the batch seasonality. The main state variables included in this simulation are: reactor temperature, biomass volume, pH, biomass concentration, substrate concentration, penicillin concentration, dissolved oxygen concentration, and carbon dioxide concentration. The full model is provided in the Appendix. The ODE23t stiff solver provided in MATLAB is used to simulate the process.

The general process flow diagram of the penicillin reactor is shown in Figure 10. In the initial growth phase, the fermenter is initiated with a small volume of bacteria culture and substrate (glucose and oxygen). The reactor temperature is kept at 298 K (room temperature) to initiate the growth of the Penicillin-producing bacteria. After approximately 50 hours, the initial substrate is depleted; the reactor then enters the penicillin production phase. In this phase, the fermenter is operated in semi-batch mode. Continuous flows of substrate (through the substrate tank) and oxygen (through the aerator in the bottom of the fermenter) promote the production of penicillin, carbon dioxide and other by-products. Furthermore, two PID controllers regulate the acidity level (pH ~5) and temperature (298 K) of the fermenter to promote penicillin production.

Figure 11 plots the resulting trajectory of a single batch. The sampling frequency is set to once every 30 minutes, mimicking that of real plant data. Output variables of the simulation include agitator power, aeration rate, substrate flow rate, cooling water flow rate, heat generation. These variables are either manipulated inputs, or measured variables of the PID control loop.

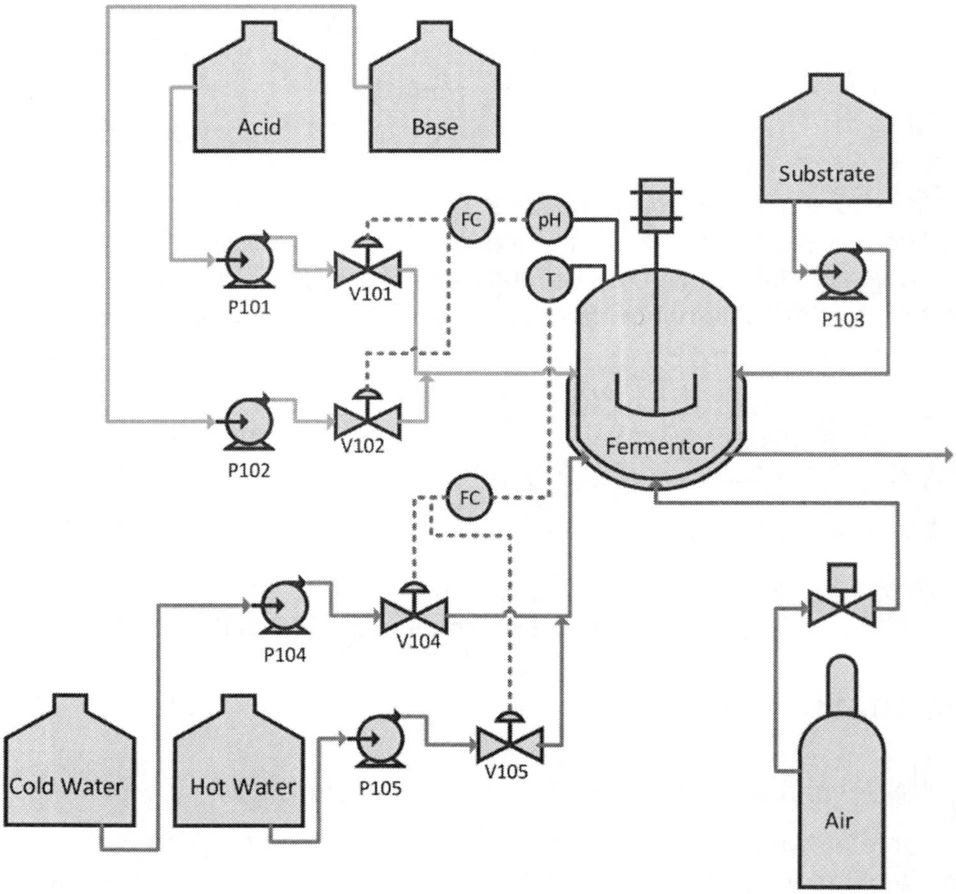

Figure 10 Process flow diagram of the batch-fed penicillin process

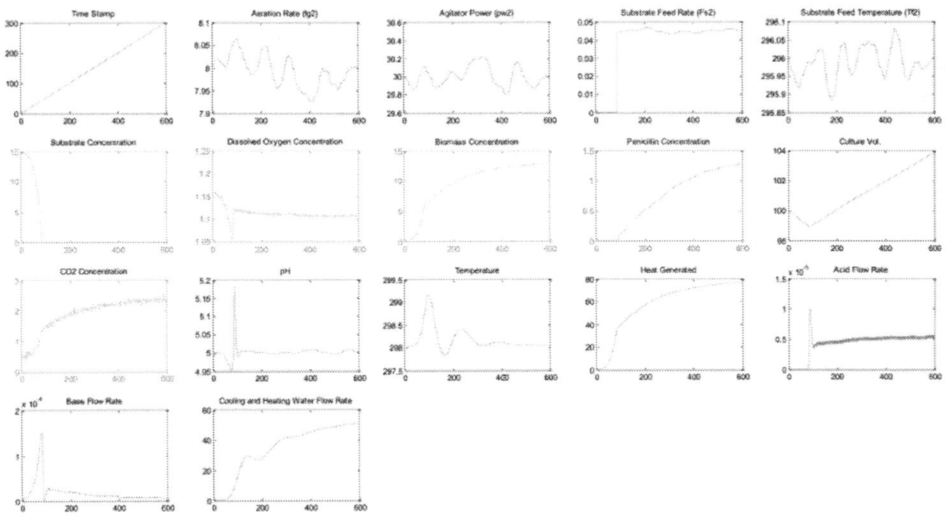

Figure 11 Example batch trajectories from a simulation run of the penicillin reactor

3.2. Simulation Conditions

To introduce process seasonality, the specific growth rate $\boldsymbol{\mu}$ in the biomass growth equation (see Equation (7)) is configured to follow a first-order exponential decay with a time constant set to 150 batches. The decay of the specific growth rate $\boldsymbol{\mu}$ starts from the nominal value of 0.092 and ends with the terminal value of 0.048. As shown in Figure 12, scheduled maintenance (which occurs once every 100 batch cycles) of the reactor resets the specific growth rate back to its nominal value.

$$\frac{dX}{dt} = \mu X - \frac{X}{V}\frac{dV}{dt} \tag{7}$$

Table 1 lists the simulation conditions for our current study. A total of 600 batches with seasonality corresponding to six cycles in Figure 12 are simulated, out of which 400 batches are used for training, and the remaining 200 are used for testing. All the data goes through batch-wise unfolding as shown in Figure 13, where 3D batch data are flattened batch-wise into a 2D matrix. After unfolding, data from the ith batch are squeezed into the ith row of the matrix, and such a row is comprised of values of variable j measured at time k. After scaling and mean centering of the flattened matrix, a lag of 10 batches is used in matrix reformatting for supervised model building. In other words, we use the prior 10 batches of data to forecast the next batch.

All the time series forecasting models are built using Python. These includes the sequence-to-sequence (seq2seq), support vector machine (SVM) regression, and partial least squares (PLS) models. The libraries used include tensor flow, sklearn, pandas, and numpy.

Table 1 Simulation conditions for batch data generation

Case	Training	Testing
	400 batches with seasonality	200 batches with seasonality

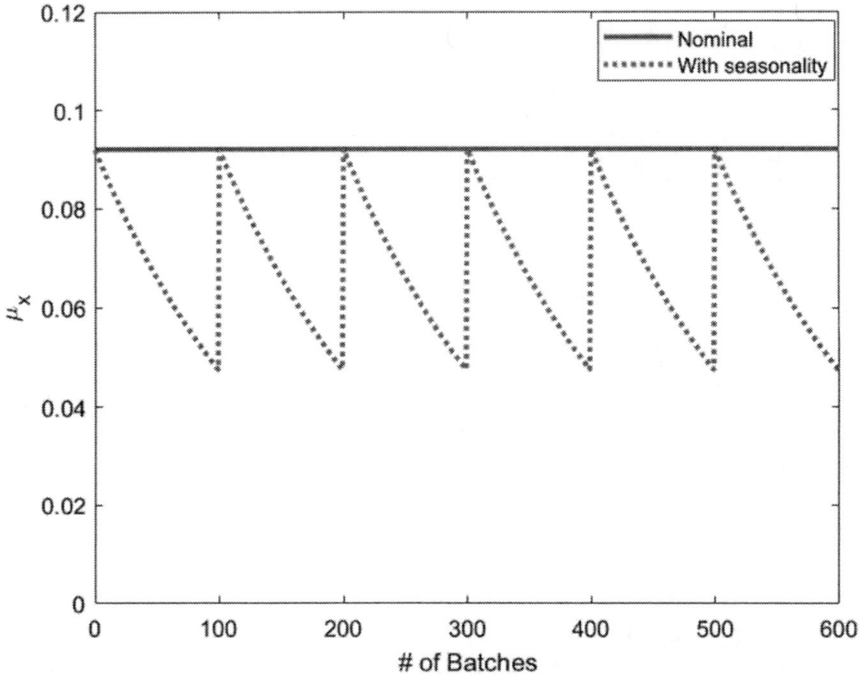

Figure 12 Seasonality in specific growth rate of bacteria culture over six cycles (decay and reset)

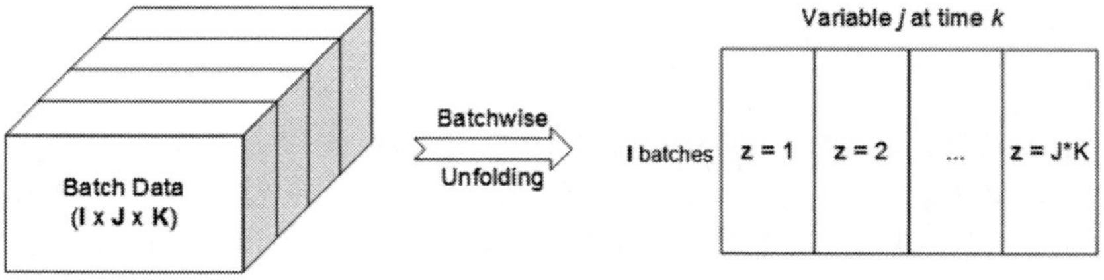

Figure 13 Batch-wise unfolding scheme in time series forecasting

4. Results and Discussions

In this study, we use the root mean square error of forecasting calculated in Equation (8) to evaluate the forecasting performance. The associated results are listed in Table 2. It is observed that the seq2seq model outperforms the other two supervised learning methods in terms of forecast accuracy of batch seasonality.

$$RMSEF = \sqrt{\frac{\sum_{i=1}^{n}(y_i - \hat{y}_i)^2}{n}} \quad (8)$$

Table 2 RMSEF values of forecasting models

Forecasting models	RMSEF
Seq2seq	3.41
SVM	4.40
PLS	4.46

The actual forecasting results are illustrated in Figures 14~16. It is observed that the nonlinear models including the seq2seq and SVM obtain smoother forecasting results than the linear PLS model. In addition, the seq2seq model excels in forecasting the exact batch where the production of penicillin rises abruptly due to reset of the bacterial growth rate.

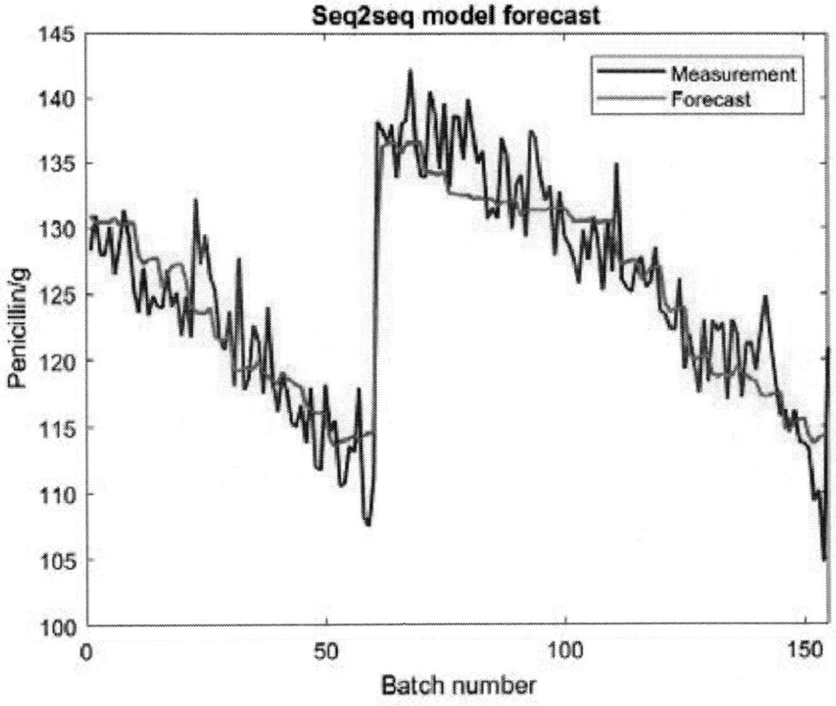

Figure 14 Seq2seq model forecasting results

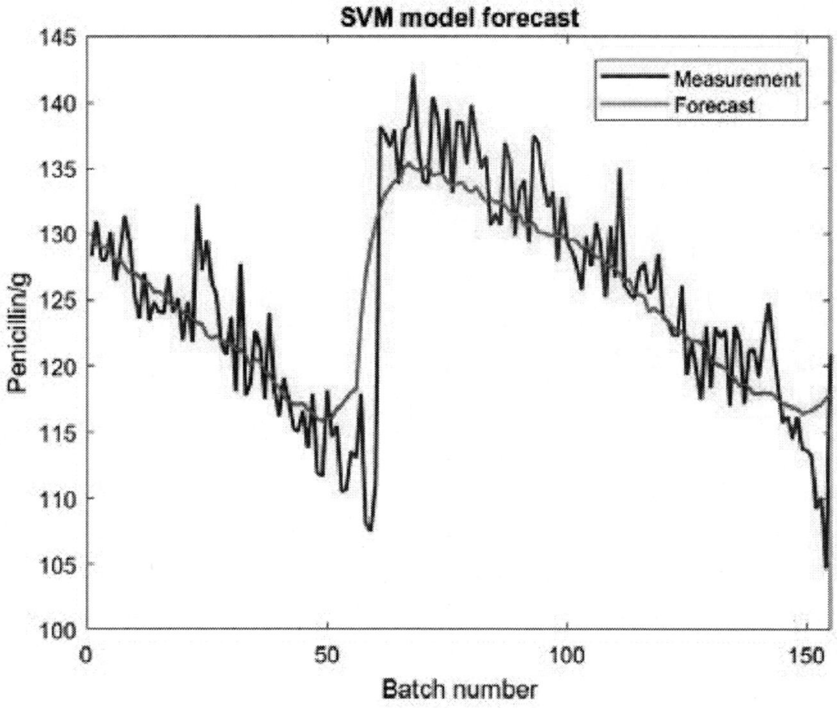

Figure 15 SVM model forecasting results

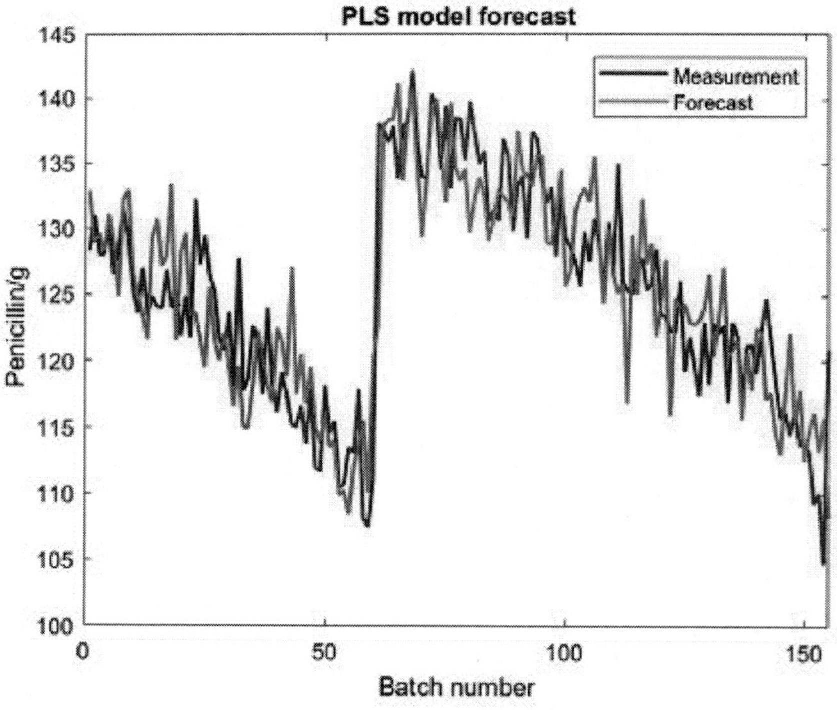

Figure 16 PLS model forecasting results

5. Conclusions and future work

We have demonstrated that the sequence to sequence learning framework can be used to overcome the challenge of batch seasonality forecasting. We built a time series forecasting model in Tensor Flow, and tested it on the data generated from the benchmark penicillin bio-reactor simulator. The model successfully forecast the cyclic change of the penicillin production rate with the periodic decay of the biomass growth rate. Such results illustrate the successful application of data analytics to wrestle with real problems in chemical manufacturing within the digital transformation era.

In the future, we will test such a seq2seq framework on more industrial data and compare its performance with other time series forecasting methods.

References

[1] C. Clarotti, A. Lannoy, S. Odin, and H. Procaccia, "Detection of equipment aging and determination of the efficiency of a corrective measure," *Reliab. Eng. Syst. Saf.*, vol. 84, no. 1, pp. 57–64, Apr. 2004.

[2] M. Ji, Y. He, and T. C. E. Cheng, "Single-machine scheduling with periodic maintenance to minimize makespan," *Comput. Oper. Res.*, vol. 34, no. 6, pp. 1764–1770, Jun. 2007.

[3] B. W. Wojciechowski, "A theoretical treatment of catalyst decay," *Can. J. Chem. Eng.*, vol. 46, no. 1, pp. 48–52, 1968.

[4] H. Müller-Steinhagen, *Heat exchanger fouling: mitigation and cleaning techniques.* IChemE, 2000.

[5] S. J. Qin, "Process Data Analytics in the Era of Big Data," *AIChE J.*, vol. 60, no. 9, pp. 3092–3100, 2014.

[6] G. E. P. Box, G. M. Jenkins, G. C. Reinsel, and G. M. Ljung, *Time series analysis: forecasting and control.* John Wiley & Sons, 2015.

[7] Y. Bengio, P. Simard, P. Frasconi, and others, "Learning long-term dependencies with gradient descent is difficult," *IEEE Trans. neural networks*, vol. 5, no. 2, pp. 157–166, 1994.

[8] S. Hochreiter and J. Schmidhuber, "Long short-term memory," *Neural Comput.*, vol. 9, no. 8, pp. 1735–1780, 1997.

[9] G. Birol, C. Ündey, S. J. Parulekar, and A. Çınar, "A morphologically structured model for penicillin production," *Biotechnol. Bioeng.*, vol. 77, no. 5, pp. 538–552, 2002.

[10] G. Birol, C. Ündey, and A. Cinar, "A modular simulation package for fed-batch fermentation: penicillin production," *Comput. Chem. Eng.*, vol. 26, no. 11, pp. 1553–1565, Nov. 2002.

Appendix – Penicillin Batch Reactor Model

Modeling Equations

The modeling equations and associated parameters and constants are taken from Birol et al. [10] and are simply provided here for completeness and convenience.

Biomass Growth

$$\frac{dX}{dt} = mX - \frac{X}{V}\frac{dV}{dt}$$

where μ is the specific growth rate and contains the effects of environmental variables (pH and Temperature) as well as carbon source (glucose) and oxygen in its kinetic expression.

Penicillin Production:

$$\frac{dP}{dt} = m_{pp}X - KP - \frac{P}{V}\frac{dV}{dt}$$

where μ_{pp} is the specific penicillin production rate containing biomass, carbon source (glucose) and oxygen concentrations in its kinetic expression. Here, the hydrolysis of penicillin is also considered and expressed as a first order rate expression with a rate constant K.

Substrate Utilization:

Glucose:

$$\frac{dS}{dt} = - \frac{m}{Y_{xs}}X - \frac{m_{pp}}{Y_{ps}}X - m_X X + F - \frac{S}{V}\frac{dV}{dt}$$

Oxygen:

$$\frac{dC_L}{dt} = - \frac{m}{Y_{xo}}X - \frac{m_{pp}}{Y_{po}}X - m_o X + K_{la}\left(C_L^* - C_L\right) - \frac{C_L}{V}\frac{dV}{dt}$$

Here, K_{la} is taken to be a function of agitator power input and flow rate of oxygen as suggested by Bailey and Ollis (1986).

Carbon Dioxide Production:

$$\frac{dCO_2}{dt} = a_1\frac{dX}{dt} + a_2 X + a_3$$

The rate expression for CO_2 has been taken from the earlier work of Montague et al. (1986).

Effect of pH

$$\frac{d\left[H^+\right]}{dt} = \gamma\left(\mu X - \frac{FX}{V}\right) + \left[\frac{-B+\sqrt{\left(B^2+4\times10^{-14}\right)}}{2} - \left[H^+\right]\right]\frac{1}{\Delta t}$$

Where $B = \frac{\left[10^{-14}/\left[H^+\right]-\left[H^+\right]\right]V - C_{a/b}\left(F_a + F_b\right)\Delta t}{V + \left(F_a + F_b\right)\Delta t}$, F_a and F_b represents acid and base flow rates in l/h, respectively.

Volume Change

$$\frac{dV}{dt} = F + F_{a/b} - F_{loss}$$

where $F_{a/b}$ is the flow of acid/base, F_{loss} accounts for the evaporative losses during fermentation: $F_{loss} = V\lambda\left(e^{5(T-T_o)/(T_v-T_o)} - 1\right)$

Heat of Reaction

$$\frac{dQ_{rxn}}{dt} = r_{q1}\frac{dX}{dt}V + r_{q2}XV$$

where *r* is the yield coefficient associated with the exothermic microbial reactions.

Energy Balance

$$\frac{dT}{dt} = \frac{F}{s_f}\left(T_f - T\right) + \frac{1}{V\rho c_p} \times \left[Q_{rxn} - \frac{aF_c^{b+1}}{F_c + \left(aF_c^b / 2\rho_c c_{pc}\right)}\right]$$

PID controller

A discretized PID control law has been used in both the temperature and pH controls. E_N denotes the error at time step N and E_{N-1} denotes the error at time step N-1.

$$\Delta MV_N = K_c\left(E_N - E_{N-1} + \frac{\Delta t}{\tau_1}E_N - \frac{\tau_d}{\Delta t} \times \left(CV_N - 2CV_{N-1} + CV_{N-2}\right)\right)MV_N$$

Overall, there are total of 15 differential equations which are solved simultaneously. All the parameters are taken from literature or assigned values which are physically meaningful in the absence of available literature. Because most of the parameters are strain-specific and are also affected by the nature of the substrate and the environmental conditions like pH and temperature, be aware that the parameters used in this simulator are average values rather than being specific to a certain strain or an environmental condition.

Initial Conditions, Kinetic and Controller Parameters

Initial conditions	
Substrate concentration: S (g/l)	15
Dissolved oxygen concentration: C_L ($= C_L^*$ at saturation) (g/l)	1.16
Biomass concentration: X (g/l)	0.1
Penicillin concentration: P (g/l)	0
Culture volume: V (l)	100
Carbon dioxide concentration: CO_2 (mmol/l)	0.5
Hydrogen ion concentration: $[H^+]$ (mol/l)	$10^{-5.1}$
Temperature: T (K)	297
Heat generation: Q_{rxn} (cal)	0
Kinetic parameters and variables	
Feed substrate concentration: s_f (g/l)	600
Feed flow rate of substrate: F (l/h)	
Feed temperature of substrate: T_f (K)	298
Yield constant: $Y_{x/s}$ (g biomass/g glucose)	0.45
Yield constant: $Y_{x/o}$ (g biomass/g oxygen)	0.04
Yield constant: $Y_{p/s}$ (g penicillin/g glucose)	0.90
Yield constant: $Y_{p/o}$ (g penicillin/g oxygen)	0.20
Constant: K_1 (mol/l)	10^{-10}
Constant: K_2 (mol/l)	7×10^{-5}
Maintenance coefficient on substrate: m_x (per h)	0.014
Maintenance coefficient on oxygen: m_o (per h)	0.467
Constant relating CO_2 to growth: α_1 (mmol CO_2/g biomass)	0.143
Constant relating CO_2 to maintenance energy: α_2 (mmol CO_2/g biomass h)	4×10^{-7}
Constant relating CO_2 to penicillin production α_3 (mmol CO_2/l h)	10^{-4}
Maximum specific growth rate: μ_x (per h)	0.092
Contois saturation constant: K_x (g/l)	0.15
Oxygen limitation constant: K_{ox}, K_{op} (no limitation)	0
Oxygen limitation constant: K_{ox}, K_{op} (with limitation)	2×10^{-2}, 5×10^{-4}
Specific rate of penicillin production: μ_p (per h)	0.005
Inhibition constant: K_p (g/l)	0.0002
Inhibition constant for product formation: K_I (g/l)	0.10
Constant: p	3
Penicillin hydrolysis rate constant: K (per h)	0.04
Arrhenius constant for growth: k_g	7×10^3
Activation energy for growth: E_g (cal/mol)	5100
Arrhenius constant for cell death: k_d	10^{33}
Activation energy for cell death: E_d (cal/mol)	50000
Density × heat capacity of medium: $\rho\ C_p$ (per l °C)	1/1500
Density × heat capacity of cooling liquid: $\rho_c\ C_{pc}$ (per l °C)	1/2000
Yield of heat generation: r_{q_1} (cal/g biomass)	60
Constant in heat generation: r_{q_2} (cal/g biomass h)	1.6783×10^{-4}
Heat transfer coefficient of cooling/heating liquid: a (cal/h °C)	1000
Cooling water flow rate: F_c (l/h)	
Constant: b	0.60
Constants in K_{la}: α, β	70, 0.4
Constant in F_{loss}: λ (per h)	2.5×10^{-4}
Proportionality constant: γ (mol $[H^+]$/g biomass)	10^{-5}

Controller parameters (PID)	
pH: (base) K_c, τ_I: (h), τ_d: (h)	8×10^{-4}, 4.2, 0.2625
pH: (acid) K_c, τ_I: (h), τ_d: (h)	1×10^{-4}, 8.4, 0.125
Temperature: (cooling) K_c, τ_I: (h), τ_d: (h)	70, 0.5, 1.6
Temperature: (heating) K_c, τ_I: (h), τ_d: (h)	5, 0.8, 0.05

554324 Text Mining and Analytics (poster)

Text mining seeks to extract information and insights from textual data. In particular the distillation and summary of documents is of increasing commercial interest, from news media to patents to maintenance logs.

Application areas include business analysis, scientific discovery, predictive maintenance, marketing and communications, litigation, and cybersecurity, among others.

Basic workflow elements will be reviewed in this talk, including: definition of the corpus (which may involve the pre-step of converting images to text), preprocessing, visualization, filtering, statistical analysis, word embeddings, and classification.

Some more advanced subjects, for example topic modeling and sentiment analysis, will be introduced and described.

Demonstrations of functionality will be presented using MATLAB.

551750 Applying Optimization Method to Deliver Proactive Pricing on Improving Margin Beyond a Cost Plus Strategy

The Dow Chemical Company is a leading chemical manufacturer in the world. Dow manufactures plastics, chemicals, and agriculture products. With a presence in over 150 countries, it employs over 50,000 people worldwide. Price is a function of volume, cost, previous price and external factors. We have provided a dynamic advanced analytical engine that can be coupled with a Product Manager's current process to maximize margin potential. This tool utilizes Price-volume elasticity while including capacity constraints and necessary external variables defined by the BU product organization.

550013 An Holistic Approach for Batch Analysis and Monitoring (Poster)

ABB Ability^{TM} BatchInsight identifies and monitors anomalies in the batch process. It helps process specialists and plant operators detect and resolve abnormalities by conducting batch-level analysis.

Using time-level analysis, it recognizes the time of occurrence of the deviation and isolates the process variables that contributed to the deviation, enabling effective and timely corrective action.

The big data algorithm empowered batch monitor supervises continuously the running batch process. It enriches a normal PV monitor with additional features like, similarity-based process forecasts, adaptive "golden batch" process corridor, batch-phase change recognition & visualization, and operator decision support after early warning.

549714 Best Big Data Analytics Platform for Manufacturers (Poster)

Artificial Intelligence, Machine Learning, and the Industrial Internet of Things are still the hot topics in the big data analytics world. Manufacturers have been generating data off the line for decades, so how can manufacturers get the most of these new advancements in analytics? The combination and convergence of new hardware infrastructures have allowed for the advancement of analytics like artificial intelligence to move from research out into the business world. Analytic insights need to be deployed into production and actions need to be taken sometimes in real-time to truly realize the benefits of all these advancements. Come see how the power of the SAS platform with it's support of the entire analytics lifecycle: Data, Discovery, and Deployment provides the best big data analytics platform for manufacturers to leverage all sorts of analytics both in batch and in real-time.

551864 Microsoft - Empowering the Chemical Industry (Poster)

Chemical companies are revisiting their own business processes and technology approaches as competitors and partners start to employ 'digitalized' business processes and exploit the increasing convergence between OT and IT on the plant floor, to connect the enterprise as a whole to the extended supply chain and throughout the ecosystem.

In this session you will learn how technology empowers workforce with devices, while AI-powered productivity tools enable them to work more effectively, manage manufacturing complexity, and collaborate specifically to their business and operational needs. Understand Microsoft approach to Firstline Worker including Mixed Reality with introduction to Microsoft Remote Assist and Layout.

551860 Business Value, Work Process, Analytics and Data in the Chemical Industry (Poster)

This presentation will begin with a discussion of the major categories of business value that can be unlocked by applying analytics to production activities, including asset reliability, production optimization, energy efficiency optimization and safety performance improvement. It will then explore how value is realized through transformation of work processes to identify several key requirements of digitalization programs and analytics, suggesting answers questions such as:

- Should I use the cloud for analytics?
- Is my process and predictive technology data sufficient to support analytics?
- Do I need to use artificial intelligence?
- How important are metadata (metadata are plural!)?

The presentation will include a demonstration of first-principles-based analytics approach applied to early detection of asset failure in support of a top quartile maintenance work management process.

551856 Prognostic and Health Management Using Digital Twin (Poster)

Digital Twin is a digital representation of your real-world device or operation that mirrors actual equipment and operating conditions. Leveraging this technology for your equipment's predictive health and operational optimization can significantly improve your company's EBITDA.

However, creating digital twins for your assets continues to be a costly affair. A typical refinery or plant has hundreds of assets. Modeling them in a robust manner is a herculean task. We will present a methodology driven by AI that works with variations in model types within a class of assets but also across different types of assets. As a result, you can rapidly scale analytics across your organization delivering extraordinary savings to your company's bottom line. Our patent pending Asset Agnostic Prognostic Technology has shown the efficiency of this approach across a variety of mechanical equipment (pumps, compressors and motor bearings. DeepIQ's prognostic health management software is

1. *a) Asset Agnostic: is easily configurable for new assets*
2. *b) Source Agnostic: connects to a wide variety of time series industrial sources*
3. *c) Platform Agnostic: works natively on both on-premise or Cloud big data environments.*

550616 Using Big Data Techniques to Tackle Refining Reliability and Optimization

I think we all recognize there is journey to get to Industry 4.0 or Refining 4.0. How do we get to the smart refinery? The smart refinery includes the Internet of Things, Cloud computing, cyber-physical process, and Analytics. Teradata is working with companies on their journey to smart refineries by helping companies with their analytics through IoT and cloud computing at scale of machine learning algorithms. Teradata is providing companies with access to the power of analytics by combining their sensor data and ERP data at granular detail level to gain insights on how to improve costs and increase gross margin in their reliability and optimization processes. We will discuss the methods we apply to to bridge the OT (Operational Technology) and IT(Information Technology) worlds together to have impacts on maintenance programs, energy optimization, and operations. We will also discuss how data integration around applications, data silos , ERP systems have to helping companies not only gain insights but also have financial impact to the bottom line at scale for multiple refineries.

551534 Transforming Abundant Data into Improved Process Reliability

Today, processing facilities are beginning to effectively utilize the abundance of process and asset health data and information as part of an overall IIoT strategy. Although a digital transformation will require investments in technology, one cannot forget the investment in training staff to behave differently (improved) with the new timely information. It should be noted that some jobs will be replaced (e.g. manual data collection), new roles will emerge (e.g. analysts and planners), and other jobs will evolve to be more efficient. Properly trained operations are invaluable in this process, as they not only act on information from analyzed data, but can also make decisions on their own based on sensor data.

Abnormal operation and imminent asset failure can quickly be identified and corrected with properly trained staff to act on this new timely analyzed information. Big data and IIoT will continue to have a role with shortages of skilled resources, thus having the staff be more efficient and effective at their role with additional information and insight into operations.

This presentation will focus on the applications where big data and IIoT have improved process and asset reliability. Although the applications presented will come from oil refineries, the solutions can be applied to other continuous processing facilities.

The Digital Plant

Industrial IoT Thoughts by Timothy Holtan (tpholtan@gmail.com)

The manufacturing manager makes hundred of decisions every week. The decisions range from the current revenue - what products at what volume - to the future expense - how much research and development to stay in front of competition. Excellent decision making combines a lot a data efficiently and objectively at a cost of speed. We will discuss a business process to speed objective decision making.

Manufacturing Manager Challenges

Manufacturing managers such as refinery managers, automobile manufacturing plant managers and nuclear power plant managers make many important decisions every day. How many units to produce. When to add overtime to increase production. What is the best product mix for the plant and for the market.

The manager faces continuous change in the form of technology advancement, regulations, labor availability, competition and other internal and external forces. Manufacturing usually operates on a cycle of first planning, then executing, and then auditing the results. These three stream in parallel as the manager continuously plans the future, executes and present and audits the past.

New Decisions, Legacy Solutions

The best manufacturing manager uses all of the tools available from ERP systems and marketing demand planning solutions and Enterprise Asset Management to make the best decisions to ensure the best future state. Managers use a familiar format to assess their performance. The variance report compares actual results to the plan and measures the difference as success or failure.

The variance report and manufacturing planning unquestionably delivers proven value. The process pushes against the limits of human execution and forecasting. Often the lens of the variance report process exposes the participant's biases. Consider a negative variance in contribution margin as an example of how these biases creep into the planning and execution process. Some managers favor top line growth and others favor cost cutting.

How much extra value could the manufacturing manager gain when he looked beyond personal bias and the staff's collective experience? What if technology offered a cold eye review of every variance and played out what if scenarios 90 days ahead in the face of raw data?

Artificial Intelligence Offers an Alternative

The rise of rapidly scaling artificial intelligence solutions offers the manufacturing manager a solution to gain a third column in the variance reports in addition to actual and plan - the prediction.

New artificial intelligence solutions demonstrate step change capabilities to combine historical data and known future data with both deterministic and nondeterministic relationships to quantify probabilistic outcomes.

Point predictive AI solutions have demonstrated profit improvements in the order of 25% to 100%. Documented examples include GE Digital SmartSignal predictive maintenance solution in the power generation industry and IBM Maximo for maintenance planning. These solutions currently operate within the function of the manufacturing process as a form of human assist and automation. For example, AI based predictive maintenance solutions in the maintenance department offer insights which help shift unplanned maintenance to planned. AI traffic solutions in the logistics department reduce transportation costs. Often a data science team - either a external vendor or an internal function - ensures a successful deployment and ongoing sustain and maintain responsibilities remain in the department.

Historical precedent suggests the AI solutions will gain even more value as the solutions comprehend more and more of the functional data. For example, as process control expanded from single loops to the multivariate advanced control to process unit constraint control value increases. Similarly logistics and production planning expanded from local into plant wide functions and transformed into the economics and planning department and focused their efforts on the mastery of linear programming solutions.

As these solutions grew in value, they specialized into separate functions beyond the original department. The process control department grew out of the operations department to provide a continued focus on utilizing the process control solutions to add value and to ensure advanced process control solutions broke out of the boundaries of the operations department. For example advanced controls could maximize objective functions across upstream and downstream processes which usually operate in data silos.

After the Information Silos

In the digital transformation parlance, this phenomena is known as breaking down information silos. Before economics and value cause AI solutions to break these silos, manufacturing managers will anticipate and accelerate AI adoption and follow the tradition of economics and planning and advanced control and form the Prediction Department

The Prediction Department will organize to use the AI solutions to continuously generate probabilistic data driven predictions for every parameter in the monthly, quarterly and annual variance report.

The manufacturing manager will derive value from the prediction department in three ways.

1. **People and Process.** The Prediction Department solves the problem of what to do with the people as the data silos fail. The manpower associated with maintaining the point solutions and departmental databases dedicated to local Optimizatlon shift to broader coverage to achieve global optimization.

2. **Sustain and Maintain**. Second, the Prediction Department solves both the knowledge transfer and the sustain and maintain challenge of the new AI solutions. Shifting manufacturing expertise to the Prediction Department and cross training the manufacturing incumbents to sustain and maintain the AI solution will allow vendors to handover analytics and predictive functions to a receptive, expert team.
3. **Focused Number Cruncher**. Finally the manufacturing manager will gain the resource to add the prediction column to the variance report.

The Prediction department will generates more comprehensive predictions from data stretching across the organization to include future demand and future costs. AI pushes the human participant to realize opportunities beyond the limits unique history and personal bias.

Vision of Success

How will the manufacturing manager gauge the success of the operations Prediction Department?

When the Prediction Department runs the variance meeting and the focus of the discussion shifts from why the plant produced 10% less profit last month to predicting how the plant could increase profit from 10% to 15% next month, the transformation succeeded.

Sunk cost economics teaches that the future is all that matters and the Prediction Department will focus the AI to change the future.

550589 Digital Ethylene Plant: The Convergence of Process and Operational

Digital Ethylene Plant: The Convergence of Process and Operational

"Refiners are poised to adopt new technologies at a faster pace after years of incremental progress; this effort could save the industry billions of dollars." (Wood Mackenzie)

Given this important convergence of process and operational worlds in the Ethylene industry, this paper looks at how operators can integrate asset performance management (APM) solutions, with process simulation models and data analytics. This convergence allows for continuous interrogation of plant data against the process model, thus providing the ability to predict and quantify operational decisions in near real-time to manage risk and boost profitability. It will reference use cases from customers who operate in Ethylene and illustrate workflow and the outcomes achieved.

AIChE Paper Number: 77B

Harnessing Ethylene Plant Manufacturing Operations Reliability using Data Driven Analytics

Dr. Pratap Nair
President & CEO
Ingenero

Kaushal Sampat
Senior Manager, Analytics & Modeling
Ingenero

Prepared for Presentation at the 2019 Spring National Meeting
New Orleans, Louisiana, April 1 - 4, 2019

Harnessing Ethylene Plant Manufacturing Operations Reliability using Data Driven Analytics

Availability of the Ethylene plant asset is an important factor that determines the Overall Manufacturing Asset Effectiveness. It is not uncommon for advantages realized from optimization and advanced controls, to get wiped out by an unforeseen plant shut down. Retro analyses of such Reliability issues have often shown that the root cause of the issue almost always exhibits symptoms way before the event. However, capturing such symptoms before it is too late and proactively addressing the issues are critical for Asset Reliability Improvement.

The symptomatic indicators reside in the large volumes of a variety of data that is measured on a continuous basis from the manufacturing operation and archived. However, identifying such symptoms is difficult and time consuming for the unaided human mind, unless it shows up obviously on a commonly monitored parameter. Such identification would entail sifting through thousands of a wide variety of data points, collected at minute wise or higher frequencies. Several of these parameters measured are non-linearly related to one another, further complicated by different types of errors in the collected data.

Big Data Analytics techniques can be utilized to help effectively analyze thousands of minute wise tag data for an Ethylene plant, in real time, to help "find problems before they find you". This paper describes how Machine Learning Techniques have been applied to Ethylene Manufacturing to provide useful, non-intuitive insights that are usually hidden in the large volumes of data collected. When carefully integrated with fundamental process models and combined with domain knowledge, it helps identify symptoms well in advance and allows for Root Cause Analysis to be done before it precipitates as a Reliability issue. The integrated approach is essential to minimize alerts on "false positives", which is important for the sustained utilization of such an application.

The multivariate nature of these techniques makes this approach different and more effective than the traditional Statistical Process Control (SPC). Sixteen Unit operations have been identified as common areas prone to unforeseen reliability issues in the Ethylene Plant. These techniques have been effectively hooked to live plant data for effective Reliability monitoring and automatic outlier identification, with focus on the sixteen identified areas. Also, coupling this with vendor equipment performance curves and fundamental predictive models, allows for live comparison of the equipment performance with its design. Commonly occurring failures can be predicted early based on pattern recognition using failure maps and signatures. Not so commonly occurring failures are also identified with indicators that allow basic engineering analysis to predict possible reliability issues.

Case examples in Ethylene plants where these techniques have enabled early identification of possible anomalies; causes of possible failures and increased equipment effectiveness are described in this paper.

The Significance of Reliability in Process Manufacturing

Starting up and sustaining production levels consistently from an ethylene plant does not always match expectations.

The raison d' être for the ethylene plant asset is to get best possible production of ethylene, without excessive costs and without incurring safety and compliance incidents. The Overall Asset Effectiveness (OAE) is a useful metric to quantitatively indicate how the Asset is performing. Details of the OAE and its calculation are covered in an earlier paper (2). The OAE provides a measure of how well the asset is doing what it is supposed to do – produce product. The key parameters impacting the OAE are Throughput (total feed to the plant), Yield (kgs product produced/kgs feed) and Availability (on-stream time of the plant).

Throughput (availability of feed and the ability of the plant to use the available feed) and Yield (the effective conversion of feed to product and its effective separation) are important to ensure performance and several initiatives focus on improving these aspects. However, it is not unusual to see the unavailability of the asset to produce offsetting these throughput and yield improvements. The Availability is largely determined by the Reliability of equipment operation.

Figure 1 shows a case in an Ethylene plant, where the plant is operating well above design on both the Throughput and Yield. Yet its total annual production falls below what the plant was designed to produce on account of below expected Availability of the asset on account of unforeseen Reliabilty issues, because of which the plant had to take unplanned shutdowns.

Figure 1

Case Example where Reliability is shown impacting the Overall Plant Performance

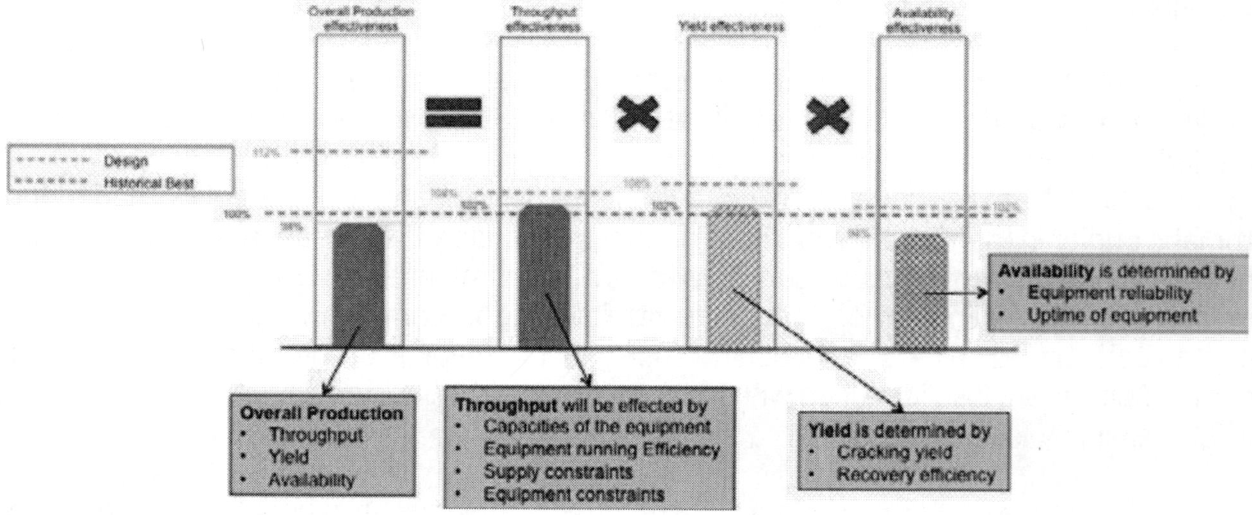

Key Challenges to be addressed

Most of the Reliability issues in a plant that requires an unplanned shutdown on account of an unforeseen issue with an equipment, upon retro analysis shows that signs of deterioration of an equipment on account of which, the plant has to be taken offline or the product quality deteriorates, was visible but hidden in the data collected from the plant and archived. However, these indicators may not be seen in the parameters the operations team normally focuses on. There are only so many parameters the human mind can observe and make sense of, at a time.

There is no dearth of data collected in most Ethylene plants, which clearly qualifies as Big Data (1,3). However, a large part of the data is just archived and not even looked at, on account of human limitations to process all the data and lack of tools to utililze the data (Figure 2).

Figure 2

Data availability not a constraint but utilization is

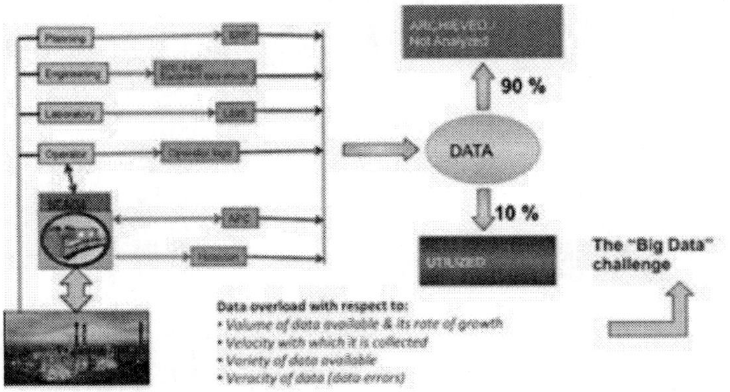

One of the key challenges that make it difficult to use the measured and stored data from the manufacturing process, as is, are Errors in data. Recent surveys have consistently shown this to be the top challenge (1), organizations face when trying to use the collected and archived data, as is. The data errors results in non-closure of the material balance around the plant. Understanding the types of errors (Figure 3) and being able to identify and weed them out of any analysis is key.

Figure 3

There are various types of errors that can skew metrics and conclusions from an analysis

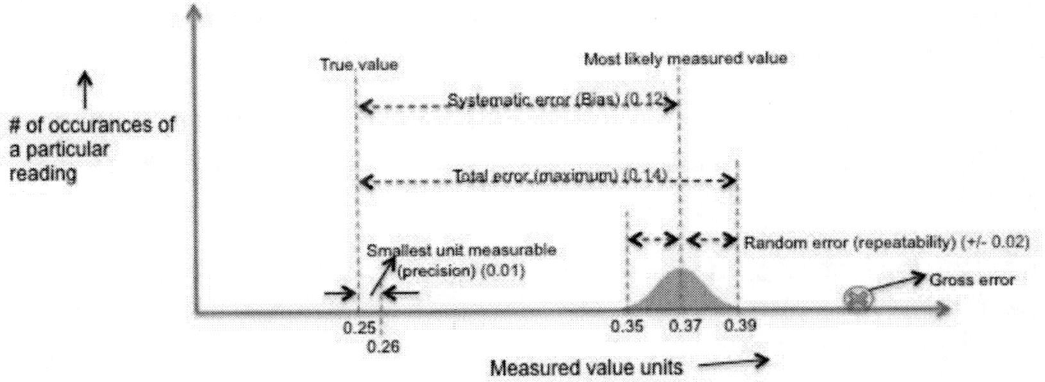

With Digitalization becoming an initiative being undertaken indsutry wide, leveraging the connectivity and affordable computing costs at feasible speeds, the veracity of the information being presented is important, since this would drive decisions. The connectivity and the viewer based dashboards from Digitalization initiatives addresses the lack of human bandwidth to process the big data. However, making sure that the information is not skewed by errors, resulting in erroneous conclusions, requires the data to be processed by analytics tools (Figure 4).

Figure 4

A basket of analytics tools and models essential for use of data when Digitalizing

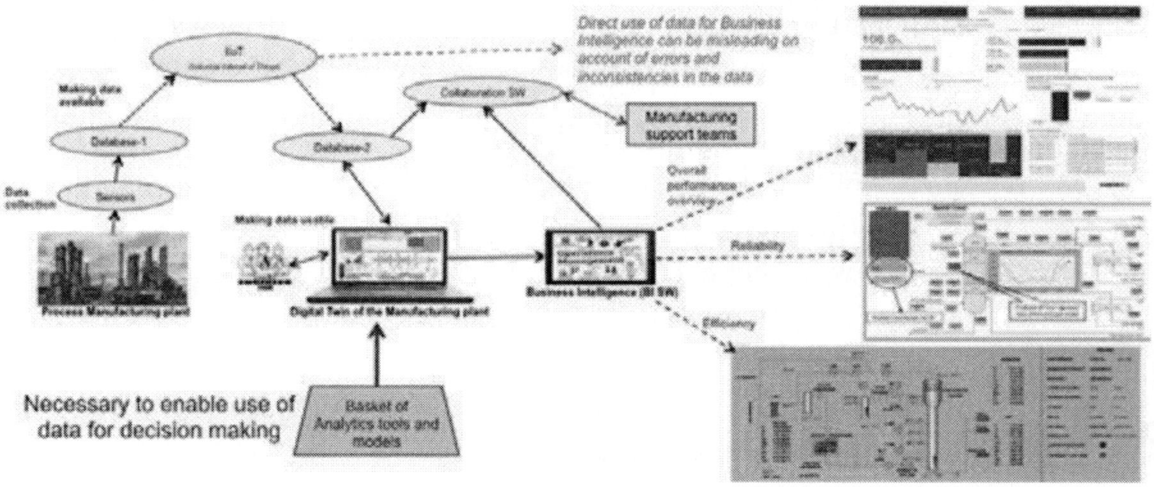

For the chemical process industry, fundamental models provide the intelligence necessary for such processing. However, use of fundamental models alone will entail a time lag between the collection of data and the conclusion. Machine Learning models, infused with intelligence (obtained from fundamental models) allows such analysis to happen near real time.

Data Analytics to improve Reliability

Reliability issues arise from equipment malfunction or quality excursions on account of changes to the health of the equipment. The key to improve Reliability, is to get an indication of the issue well in advance, so that there is sufficient time to address the potential problem or delay its ramifications till the next planned shut down, without suffering the losses on account of unplanned outages or quality excursions.

Predictive Analytics using auto adaptive Machine Learning models helps find problems before they find you with sufficient time to identify root causes and fix before an unforeseen shutdown or quality excursion (Figure 5).

A pure Data Science approach to use data from Process Manufacturing is likely to result in several false positives (false alarms). Alert fatigue from a purely data driven approach can be alarming. Studies have shown less than 5% of anomalous alerts flagged by a pure data driven Machine Learning algorithm, were actual threats – rest were false positives.

Figure 5

Early identification for improved Reliability

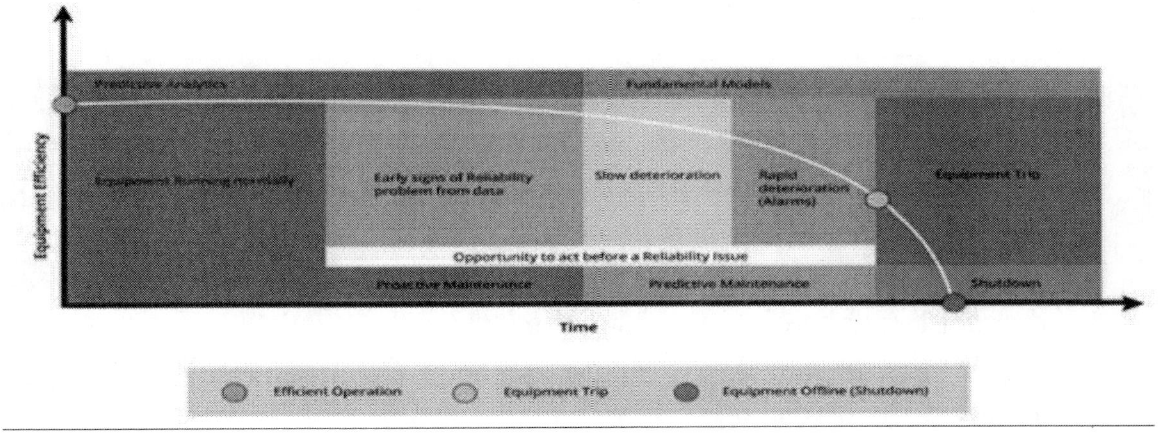

Figure 6 shows an application of utilizing Machine Learning to track equipment and flag early warning alerts, minimizing false positives.

Figure 6

Integrated solution for early identification of Reliability issues

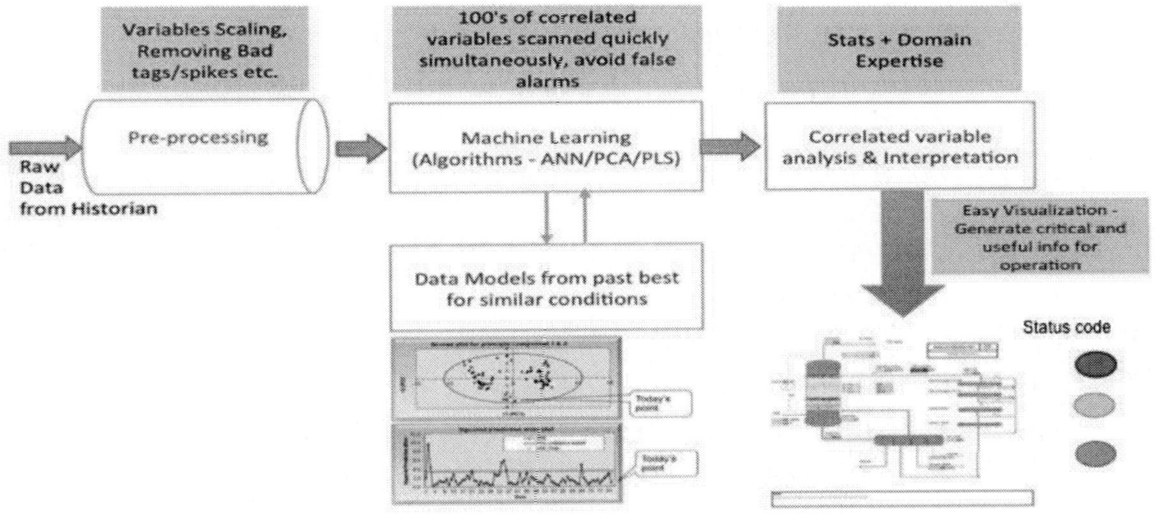

Insights we have gained through experience indicate several factors need to be considered when building Data Driven Machine Learning Models for Process Manufacturing

applications:

- ➢ Objective for which the model is being built should be clear
  - ✓ This will determine the choice of Training and Testing data used to build and validate the model
    - *For example, fouling detection models must be trained on clean condition data*
- ➢ Choice of Inputs and Outputs requires domain knowledge
  - ✓ Outputs determined by what we want to predict
  - ✓ Identifying variables that have maximum impact on predictions and whether their coefficients make directional sense from process perspective
  - ✓ The Degrees of Freedom and the control logic of every equipment have to be thought through before finalizing the input-output variables
  - ✓ Use of inferred variables as inputs instead of just measured variables greatly improves the prediction quality
    - *For example L/G, ΔT, SHC, P/E versus just the raw measurements*
- ➢ Proper filtering of training data
  - ✓ Based on purpose for which the model being built
  - ✓ Effectiveness of the Training data to enhance its applicability and reliability
    - *For example, Signal variation to noise ratio high enough*
- ➢ Combining Classification Techniques with Regression techniques (CART) can simplify the model prediction effectiveness and efficiency
  - ✓ Domain expertise aids the process in addition to classification based purely on the variable tree hierarchy
    - *For example, choose Feed type versus age of coil for furnace run length prediction*
    - *For example rather than building a non-linear model that covers all scenarios, categorize (classify) data by feed slates i.e. different models for different feed slates (9E-2P, 8E-3P, 7E-4P) versus combining all feed slate models in one*
- ➢ Provision in the model to learn on its own as equipment health changes and the model can be re-calibrated
  - ✓ Required to prevent false positives or recurring flagging of anomalies, even after it has been recorded

Machine Learning based Early Warning Tracker

An early warning tracker for Ethylene plants was developed and is being used as part of Monitoring and Tracking initiatives across multiple facilities. For the Ethylene facility sixteen key equipment areas were identified, based on historically occuring areas where Reliabilty issues are commonly observed, and monitored by the tracker:

- ➢ Furnaces
- ➢ Quench Tower
- ➢ CGC Machine
- ➢ Turbines *(CGC, PRC, ERC)*
- ➢ Cold Box
- ➢ QW loop
- ➢ Caustic Tower
- ➢ CG Dryer
- ➢ Saturator
- ➢ DOX Box
- ➢ MAPD Bed
- ➢ Condensate Stripper
- ➢ Deethanizer
- ➢ Debutanizer
- ➢ C2 Splitter
- ➢ C3 Splitter

The tracker is connected to the Historian and tracks in real time and detects anomalies early, just as it begins to see variations, not just on one specific variable but considering several related variables. Some of the features of this tracker:

- ➢ It sends automatic emails upon Outlier identification
- ➢ False Positives reduced by following the integrated solution approach as shown in Figure 6
- ➢ First Principles Models/ Vendor Performance Curves/ Column Loading Charts incorporated
- ➢ The analysis is multivariate hence different from SPC (which is typically uni dimensional)
- ➢ Identifies early signs of failure or Equipment performance degrade
- ➢ Possible causes of the outliers are identified and alerted
- ➢ Allows trending of outliers
- ➢ Easy deployment at site and linking to Historian with minimal maintenance

A snapshot of the typical output from the tracker is shown in Figure 7.

Figure 7

Early Warning Tracker output

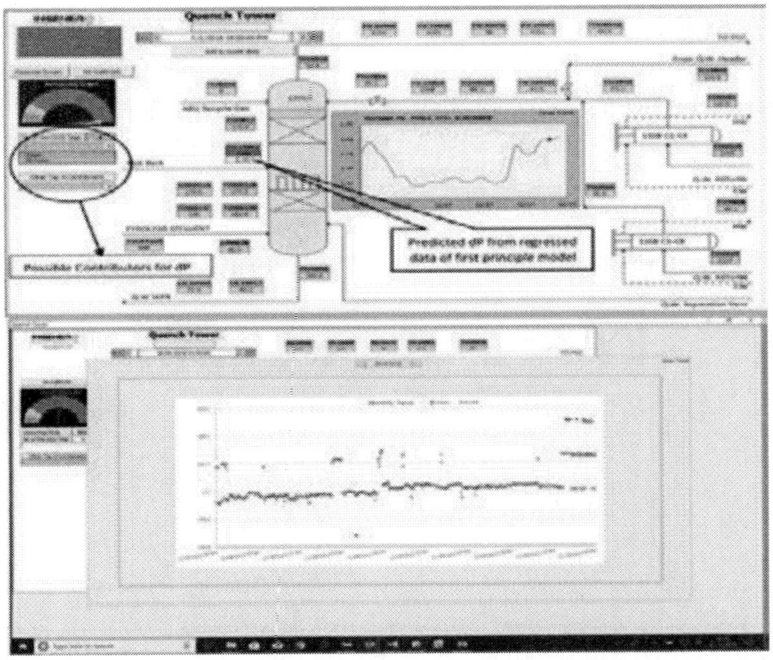

Scanning thousands of correlated variables in a typical Ethylene heat-integrated plant is not possible for the naked eye. Finding problems early helps avoid unplanned production losses and also helps in narrowing down variables for Root Cause Analysis. The automatic analysis of several correlations is only possible through machine learning models and takes and order of magnitude higher effort if it were to be done manually. The intelligence incorporated from independent runs of fundamental models and domain expertise to select appropriate inferred variables as model inputs, is essential to make the analysis results "reliable".

Equipment Reliability improvement Case Examples using the Tracker

Three case studies have been selected, to illustrate the utility of the Reliability Tracker:

1. Live Fouling detection for the C3 Splitter Reboiler
2. Live Machine Learning based operator guidance for C2 Splitter
3. Retro analysis using the Reliability Tracker for a Quench tower

Live Fouling detection for the C3 Splitter Reboiler

The C3 Splitter reboiler is heated by Quench water on the utility side. The quench water circuit is prone to fouling and this is a common occurrence on this reboiler, which then starts impacting the performance of this column making it a plant constraint.

A purely UA based calculation, which is the typical paradigm, is too simplistic and leads to many false positives. The parameters around this system are strongly correlated to each other, making simple multivariate regression unviable as a solution.

Once the Machine Learning model was well posed, it was found to successfully identify early signs of fouling in the reboiler, almost a month in advance of it starting to impact the process performance. The model is trained on clean reboiler conditions, which then serves as a multidimensional reference for "true" outlier detection, that indicates early fouling. The Quench water outlet temperature was found to be a key indicator, reflecting this comparison, to be shown on a 2-D plot, for easy visualization of deviation (Figure 8).

Figure 8

Early indication of fouling more than a month before process impact

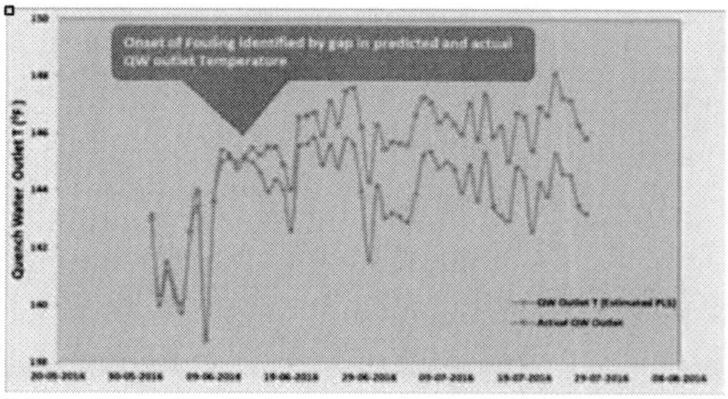

Choice of inputs and outputs, with appropriate data filters to capture the input/output variability adequately, to weed out noise and calculate "true" correlation coefficients, is required and critical (Figure 9).

Figure 9

Machine Learning Model I/O for a C3 Splitter

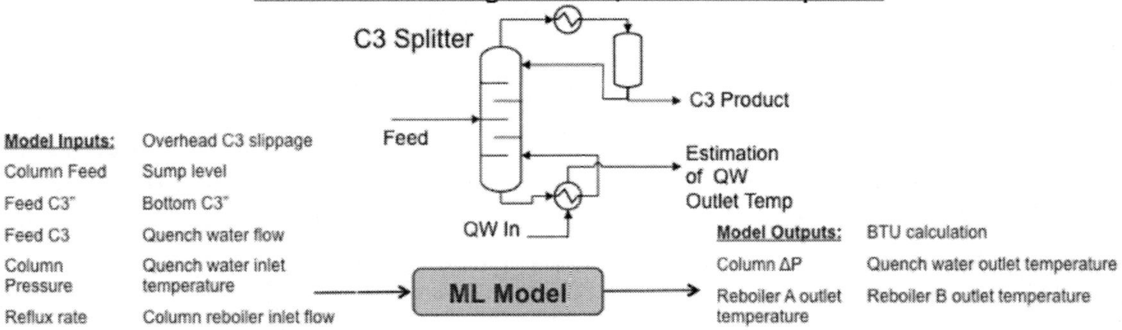

Live Machine Learning based operator guidance for C2 Splitter

The C2 Splitter Controls are often placed on manual (even with an APC) and Setpoints often end up being based on operator intuition. The Ethylene product often goes off-spec on Ethane in the Ethylene product, especially when feed rates go up.

The Reliability Tracker, by virtue of the underlying Machine Learning based model is able to provide live guidance to operators, beyond just alerting on deviations while tracking, to help avoid quality excursions, which is another issue related to asset operations Reliability in delivering not just volumes of product but also the required quality of product. Figure 10 shows an example of live recommendations of Setpoints provided to ensure on-spec product.

Figure 10

Operator Guidance for C2 Splitter from the Tracker

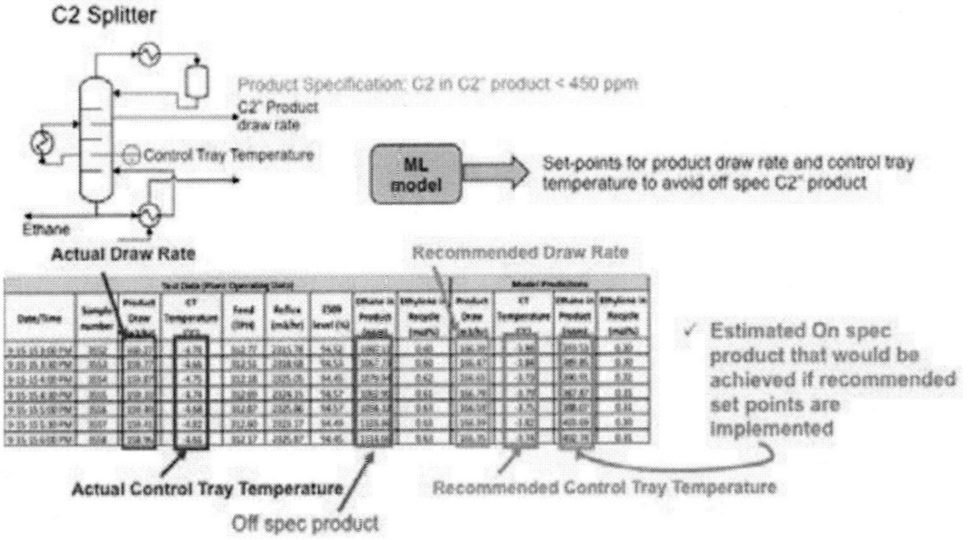

Retro Analysis using the Reliability Tracker

The Reliability Tracker was used to perform a Retro analysis of a past failure of a Quench Tower, based on historical data. The tracker was configured and then made to run through the historical data, like it was doing a live tracking of it, in a fast forward mode.

The Tracker started indicating variations several months before the failure occurred and even indicated possible root causes of the failure, as part of its flagging potential causes for the variations. The interpretation of the indicators flagged by the Tracker helped the process monitoring team easily identify fouling of the Quench tower, on account of consistent operation of the furnaces at high Propane to Ethane (P/E) ratios, without increasing the Steam to Hydrocarbon ratio, to reduce the formation of greater quantities of tar, on account of the higher heavies content in the feed. More over, the removal frequency/rate of tar from the Quench bottom was also not increased from when the plant was run at lower P/E ratios of feed.

The diagnosis would seem obvious to a process engineer monitoring the plant, once the key parameters are seen in conjunction, which may be easily missed if it is not presented or flagged by exception, as done by the Tracker. Had the Tracker been operational before the failure, the plant process monitoring team would have picked up on this quickly and fixed the root cause, thereby preventing the fouling to escalate to a stage of equipment failure and an unplanned shutdown to change the damaged column and loss of about one month of production.

Manual independent trend analysis of the historical data, post obtaining the insight from the Tracker analysis, as a cross verification, indicated that the furnaces were indeed run at higher Propane to Ethane feed ratios and lower Steam to Hydrocarbon ratios, instead of higher SHC ratios as would have been prudent (Figure 11).

Figure 11

Manual trend analysis of historical data

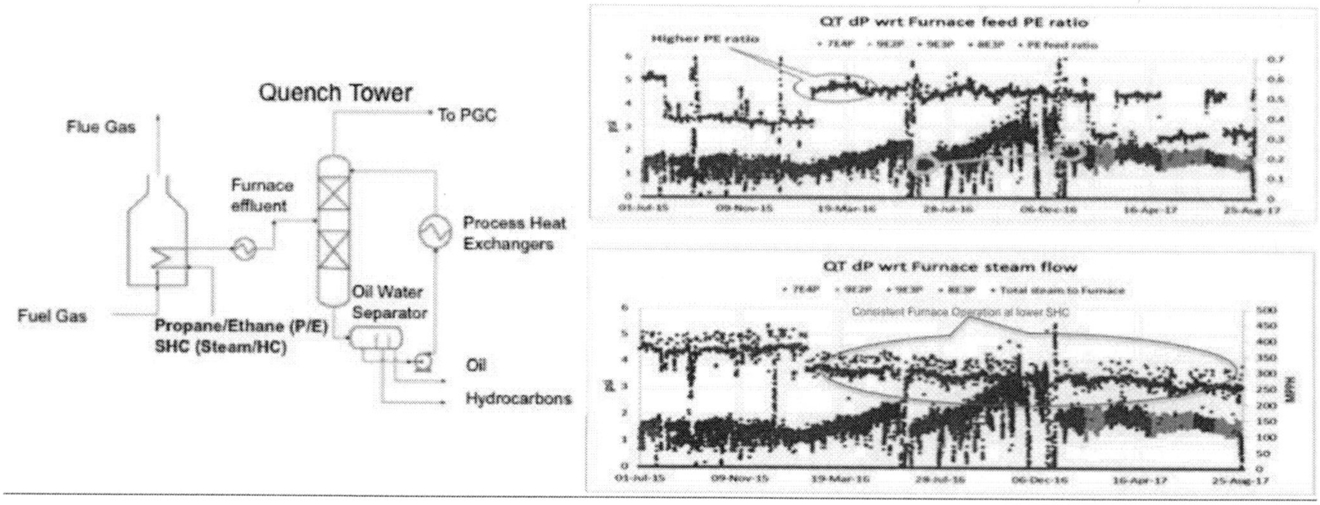

In the Quench model, predicted pressure drops from regressed data of a first principles model, was used a reference to prevent it being erroneously flagged as an outlier. This is an example of "adding intelligence from a first principles model" in the Machine Learning model (also shown in Figure 7).

Significance of domain expertise in the Reliability Tracking analysis using Machine Learning

Domain expertise is a major enabler for successful application of Machine Learning algorithms to Process Manufacturing applications. The impact of inferred variable selection on the results that can be inferred is normally significant, which requires domain knowledge:

- Machine Learning algorithms are more effective if critical "Inferred" variables used to build data driven models
- This approach gives true outliers and helps doing a quick root cause analysis versus just doing the analytics on pure data from tags leaves the interpretation to the end user
- Chances of flagging false positives are high if only raw tags are used to build data driven models
- Knowing equipment's typical pain point and inferring tags around that highlights non-intuitive insights

In the Quench Tower example above, the following selection of inferred variables made a difference versus relying on measured tags alone:

- Temperature Tags across the mid-section of the Quench Tower were replaced by Delta Temperatures across the bed
    - ✓ Uniform distribution of furnace effluent inside the packed bed should indicate minimal T gradients
    - ✓ Channeling or fouling in the bed would flag DT as outlier
    - ✓ Raw T tag may not flag as outliers
- To track relative Vapor-Liquid traffic inside the quench tower, tracking L/G ratio vs only L and G separately has multiple advantages in correlating with column ΔP
- Ratio of Total Propane to Ethane feed to the furnaces and the total overall SHC of all furnaces together is a critical inferred parameter which impacts fouling in Quench Tower

Value added analytics to the Machine Learning based Reliability Tracker

Associated analytics using trends and Historical data from the Reliability tracker adds value. Two such examples are, failure prediction and live benchmarking, which will be briefly covered as part of this paper.

<u>Failure prediction from Degradation analysis</u>

Failure prediction requires past data from the time of a similar failure. This is then used as a fingerprint for recognizing the same pattern in future amongst the thousands of data points. Often times failure data may not be available. In this case another possibility is to do a probabilistic model analysis based on tends from the Reliability Tracker. Such an analysis may be valuable, as it gives a sense on the possible time to failure and the time that one has to fix the problem or prepare for it.

This was applied successfully to a highly mechanical system for a process with eighty Coal Gasifiers (Figure 12).

Figure 12

<u>Degradation Tracker analysis for Time to Failure</u>

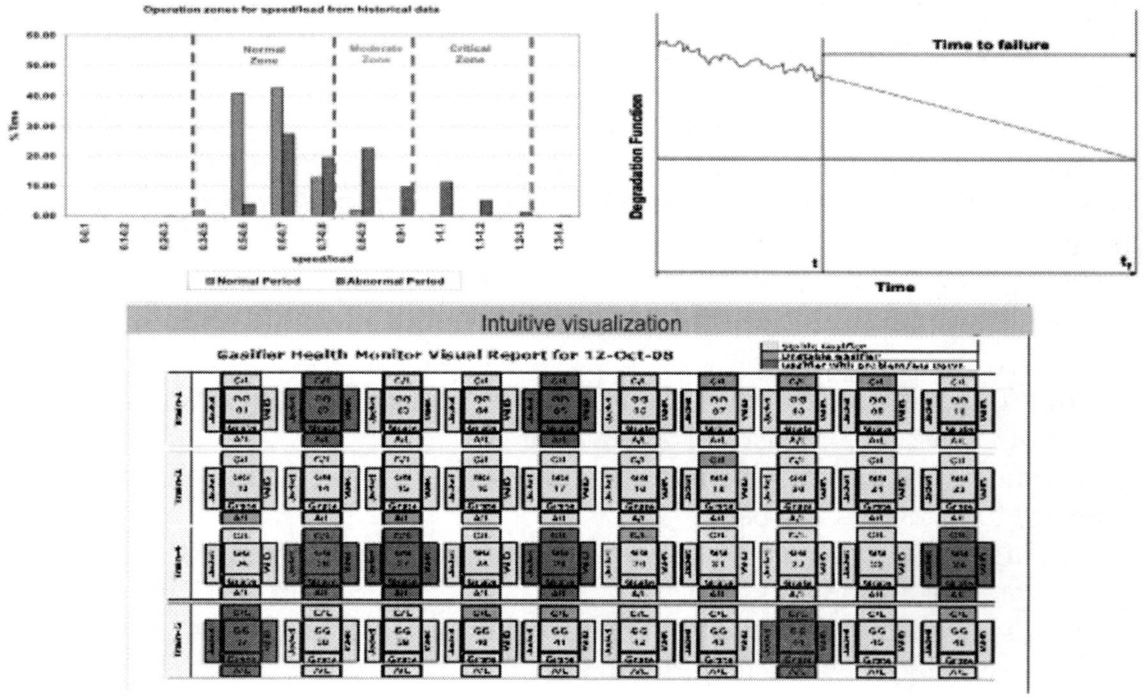

The Tracking coupled with the degradation analysis

- ➢ Enabled proactive maintenance and minimized unforeseen production capacity losses
- ➢ 7% improvement is asset availability
- ➢ 30 days alert versus just 4 days to take proactive action

This application is applicable to several scenarios in the Ethylene plant and other process manufacturing facilities:

- ➢ Component/equipment health monitoring
- ➢ Maintenance scheduling. Prioritizing maintenance within a fleet of equipment
- ➢ RCA and corrective action in mechanical and operational aspect in case of abnormal deterioration pattern
- ➢ Generalisation to any system/process
    - ✓ Reactors/Catalyst
    - ✓ Furnaces affected by coking in tubes

- ✓ Heat exchangers, condensers, after coolers
- ✓ Pumps, compressors, expanders

Live Benchmarking

Comparison with a yardstick is required, if we are to conclude on performance. Benchmarking with previous best operation and with design are two readily available yardsticks that can be used with available data, without having to run separate predictive models. Comparison to an ideal benchmark requires tuned fundamental models, which is more complex and time consuming and cannot be done in real time.

Machine Learning techniques are being used to do automatic live benchmarking of key metrics, like Yield or Production with previous Historical best and Design, for comparable conditions, from available historical data. The quick benefits that are gained from this automatic analysis would usually take several man-hours of trending and data analysis even for experienced domain experts. It saves the manual effort of interpretation derived from trending several variables together on a historian.

This analysis enables, the automatic identification of the best condition, based on a defined Performance metric that the plant has operated at, from the historical data made available to it and compare it with the current operation, for any defined equipment or section of the plant. The equipment/unit operation constraint and any possible operating change that could be readily made to close the gap, if any, between the current operation and the historical best, is identified and indicated.

The defined section of the plant could be, individual equipment, a Unit operation, the entire hot or cold side or the entire plant. The Performance metric would be the one that drives the comparison, while the result would be an explanation and/or an action to close the gap. The tool based on this analysis, ensures that the historical operating point selected is an "apples-to-apples" comparison, by matching key operating parameters of the plant from the current operation.

Such automatic benchmarking analysis when coupled with the live Reliability Tracker offers various benefits. It enables:

- ➢ Automatic Identification of Heat Exchanger Fouling
- ➢ Tracking Coil degradation and how much it is costing in terms of Yield
- ➢ Identifying faulty instrumentation automatically
- ➢ Hydrate formation identification in Cold Section
- ➢ On-the-fly comparison of Summer-Winter, Day-Night bottlenecks and how much it costs
- ➢ Early identification of Compressor degradation
- ➢ Identifying APC model drifts and tracking efficacy of APC
- ➢ Live Comparison of APC limits and Setpoints from historical data
- ➢ Tracking Furnace Differential Cracking, Furnace Efficiency and Burner Health, Decoking efficiency etc.

- ➢ Live Hydraulic tracking of Distillation Columns like Demethanizer, Deethanizer, Depropanizer, Debutanizer, C2Splitter and C3 splitter
- ➢ Automatic ERC, PRC comparison and its overall impact on C2 losses
- ➢ Tool can be configured at equipment level, section level and plant level
- ➢ Helpful for tracking plant Overall Asset Effectiveness, Equipment Availability and helps in decision making for Capex/Revamp projects
- ➢ Identifying Aging/non-performing equipment and its impact on Margin.

Figure 13 shows an example of the Deethanizer as the defined section, with the bottom ethane spec as the Performance metric. It automatically identifies and shows previous best Bottom Ethane slippage from the available history, for conditions equivalent to the current operating conditions. In this example, the gap is partly caused by C3" in Ovhd which is the constraint, but the gap can be closed by increasing Control Tray Temperature Set point.

Figure 13

Live Benchmarking for a Deethanizer

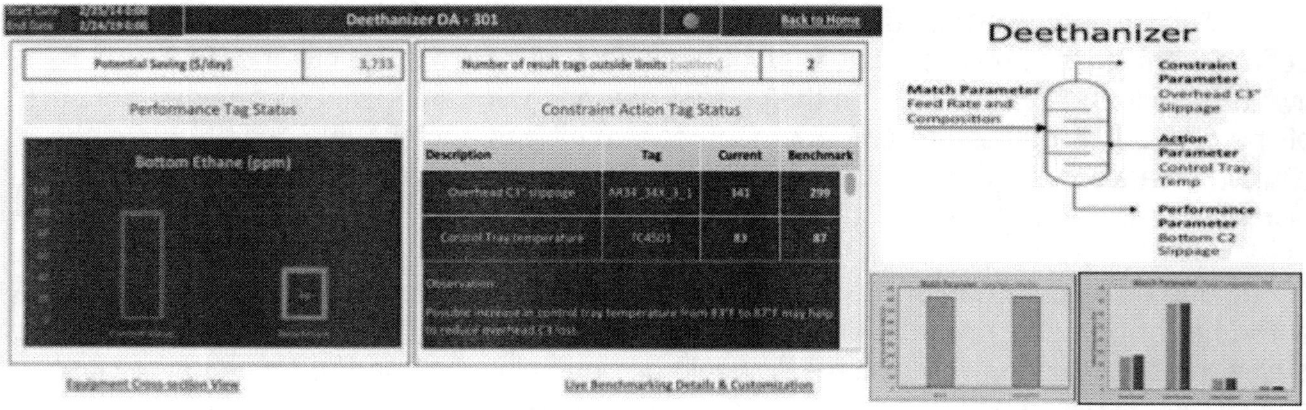

Conclusions

- ➢ Equipment Reliability and hence availability of the plant asset is one of the key parameters that impacts the overall effectiveness of the asset
- ➢ There is sufficient data available to be able to detect anomalies and potential reliability issues, albeit not readily usable in the raw form available in the historians
- ➢ Utilization of Machine Learning techniques enables us to closely track and draw out the early warning indicators, allowing us to find reliability problems before they find you and address
- ➢ Machine Learning algorithms allow closer to real time information
- ➢ Machine Learning models used by itself could result in several false positives
- ➢ Machine Learning models used in conjunction with distilled intelligence from fundamental models and domain expertise provide more robust and usable results
- ➢ Other Machine Learning techniques for failure prediction and live benchmarking offer enhanced value from a reliability tracker

References:

1) P. Nair, (March 7, 2017) "Tailoring Big Data Analytics for Optimizing Ethylene Plant Asset Performance" 2017 Spring National AICHE Meeting, Paper 2B (Ethylene Producers Conference).
2) P. Nair, (April 23, 2018), "Big Data Analytics for Improved Overall Asset Effectiveness of Ethylene plants" 2018 Spring National AICHE Meeting, Paper 169B (Ethylene Producers Conference).
3) L.F.Colegrove, M.B. Seasholtz, C. Khare., (March 2016) "Big Data – Getting Started on the Journey", CEP, pp. 41-45.

AIChE Paper Number 77c

Enabling Transfer Line Exchanger Utilization Optimization based on Advanced Data Analytics

Dr. Jörg Weidenfeller
Director RnD / Technology
SCHMIDTSCHE SCHACK | ARVOS GmbH:

Prepared for Presentation at the 2018 Spring National Meeting
New Orleans, Louisiana, April 01-04 2019

AIChE Paper Number 77c

Enabling Transfer Line Exchanger Utilization Optimization based on Advanced Data Analytics

Dr. Jörg Weidenfeller
Director RnD / Technology
SCHMIDTSCHE SCHACK | ARVOS GmbH:

Abstract:

Operation and reliability of transfer line exchangers are critical in the production of Ethylene and the economics of the plant.

The focus of the operator is on the process itself, for which standard sensor readings are monitored. Those readings up- and downstream the TLE nowadays are used to derive the condition of the heat exchanger. The interpretation of this data depends on the knowledge and experience of the individual operator. In addition regular inspections are conducted which only provide a snap-shot with limited confidence for future condition.

To support the needs of Ethylene plant operators and maintenance team the TLE functionality was further improved by applying Industrial Internet of Things technology (IIoT). In this case the TLE is equipped with additional sensors, installed fixed at measurement locations of interest for the TLE but not for the process. From the sensor readings meaningful values and equipment Key Performance Indicators (KPI's) are derived and monitored. Additionally the data are used for models which are based on OEM's domain knowledge with respect to specific thermal and mechanical design experience. In addition artificial intelligence (AI) is used to discover abnormalities in operation, to forecast the reliability and performance of the TLE and to provide diagnostics and future recommendations.

Introduction

SCHMIDTSCHE SCHACK | ARVOS GmbH, formerly known as ALSTOM Power Energy Recovery is designer and manufacturer (OEM) of transfer line exchangers (TLE) in Ethylene steam crackers and waste heat recovery systems in a number of chemical, petrochemical and metallurgical processes. The company was founded more than 100 years ago and was the first who introduced and established steam super-heaters for steam locomotives. Until then steam locomotives operated on saturated steam with an inappropriate poor efficiency. The founder of SCHMIDTSCHE SCHACK had the idea to improve this product by implementing a new technology in the existing envelope of the steam locomotive which significantly increased the efficiency of the steam process.

This company philosophy and motivation have not changed since then, but the products and applications did. More than 50 years ago, SCHMIDTSCHE SCHACK developed the transfer line exchanger (TLE) and established it in the Ethylene industry to replace the quench mode. Today, further enhancements to the TLE are being introduced continuously while keeping the existing envelope.

The operation and reliability of the transfer line exchangers are critical issues in the Ethylene process with impact on plant economics. In the transfer line exchangers the sensible heat in the cracked gas of an ethylene furnace is converted into high pressures steam (HPS). HPS is used to drive turbines, compressors and pumps in the separation train of an ethylene plant and is most important in the crackers energy system. Any loss of steam production by failure or decreased performance due to fouling of the TLE has to be compensated by HPS import from boilers requiring additional fuel gas or fuel oil with a negative effect on plant economics. High reliability and performance of the TLE's are most important, as utilization of the TLE's have a significant impact on the plant economics and require a high quality of the TLE, for maximum utilization and performance.

The status quo in Ethylene plants is that sensors and instrumentation are installed to monitor and control the production process and the process flows. From those measurement locations upstream and downstream the transfer line exchanger the status and condition of the hardware might be derived. The interpretation of these data strongly depends on the knowledge and experience of the individual operator. The real status and condition of the transfer line exchanger is obtained only during regular shut-down of the production line and inspection of the TLE. This indirect approach has uncertainties due to

- data are not collected directly at the TLE and do not describe TLE condition
- inspection during shut-down only provides a snap-shot with limited confidence for future condition

In general availability and performance are independent of each other, as a heavily fouled TLE will operate at a low performance level. Maximizing the availability and performance of the TLEs by online monitoring and predictive maintenance increases the asset utilization and will improve the plant economics significantly.

Enabling The Smart TLE

To support the needs of Ethylene plant operators SCHMIDTSCHE SCHACK started to further improve TLE's functionality by applying Industrial Internet of Things (IIoT) technology while keeping TLE's envelope. This requires a stepwise approach as shown in Figure 1. At the base level the TLE is equipped with additional sensors which will be installed during the manufacturing process in the SCHMIDTSCHE SCHACK workshop. At site the sensors will be installed after erection of the TLE. The measurement locations are at the points of interest for the TLE but not for the process. Those sensor readings are input for further evaluation and assessment of the status and condition of the TLE. The data are acquired and stored in an independent data acquisition system (DAQ) which will be installed nearby the TLE. The DAQ consists of Input/Output controllers connected to an edge-device which sends the live data via high-speed wireless network to the SCHMIDTSCHE SCHACK data storage. The complete DAQ system is enclosed in an explosive protected (ATEX) control box.

This approach ensures that the Data Control System (DCS) of the plant will not be impacted or disturbed. Optional operational data can be integrated by connecting to the DCS if the operator requests this. The SCHMIDTSCHE SCHACK data storage manages the storage and data handling. Live data are evaluated and visualized, historical data are stored, meaningful values and equipment Key Performance Indicators (KPI's) are derived and monitored.

The data are used for further models which are based on OEM's domain knowledge with respect to thermal and mechanical design. This enables diagnostic of the current TLE status. In addition artificial intelligence (AI) is used to discover abnormalities in operation, to forecast the reliability and performance of the TLE and to provide diagnostics and future recommendation service.

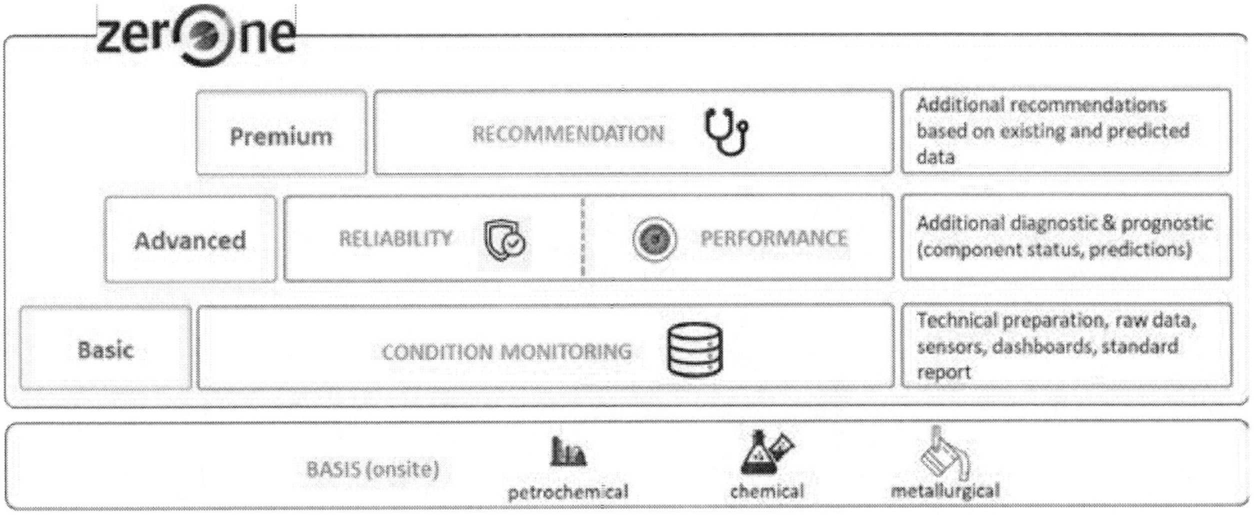

Figure 1: Stepwise approach to increase asset utilization

In summary this describes the new SCHMIDTSCHE SCHACK Zero.One platform for collection and analysis of data from TLEs. Zero.One is a cloud-based platform as a service and enables asset performance management and analytics by connecting equipment, data and people.

Smart Information Create Value

A typical view of a DCS monitor is shown in Figure 2. The data are displayed in time series and significant sensor values are marked. Although this presentation of data is well known it is difficult for the operator to identify the current status of the TLE as well as the future condition.

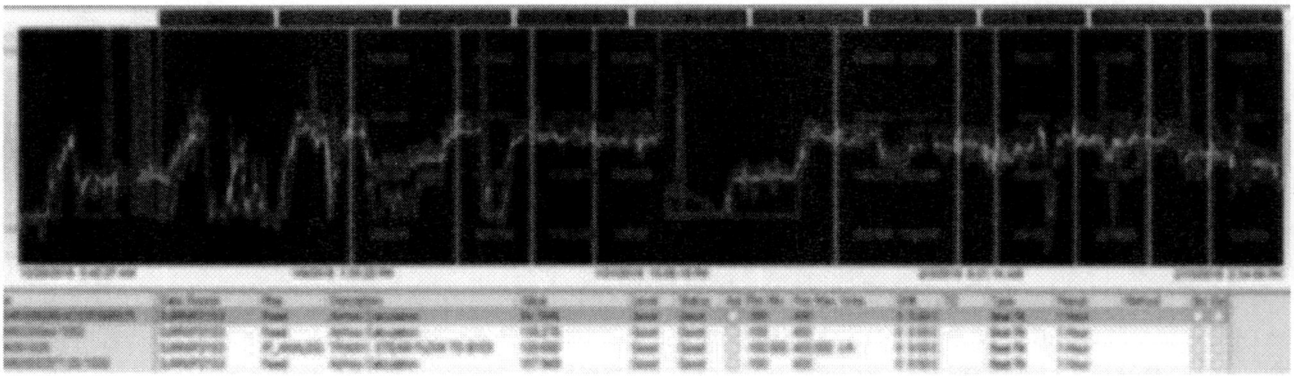

Figure 2: Typical DCS monitor view (blurred)

In contrast to that, Figure 3 shows a monthly KPI dashboard for the TLE. Several KPI dashboards for each TLE installed at the plant are aggregated to a single plant wide KPI dashboard. This information ensures a quick overview about the current and future condition of all TLEs by accessing the web-based platform mentioned above.

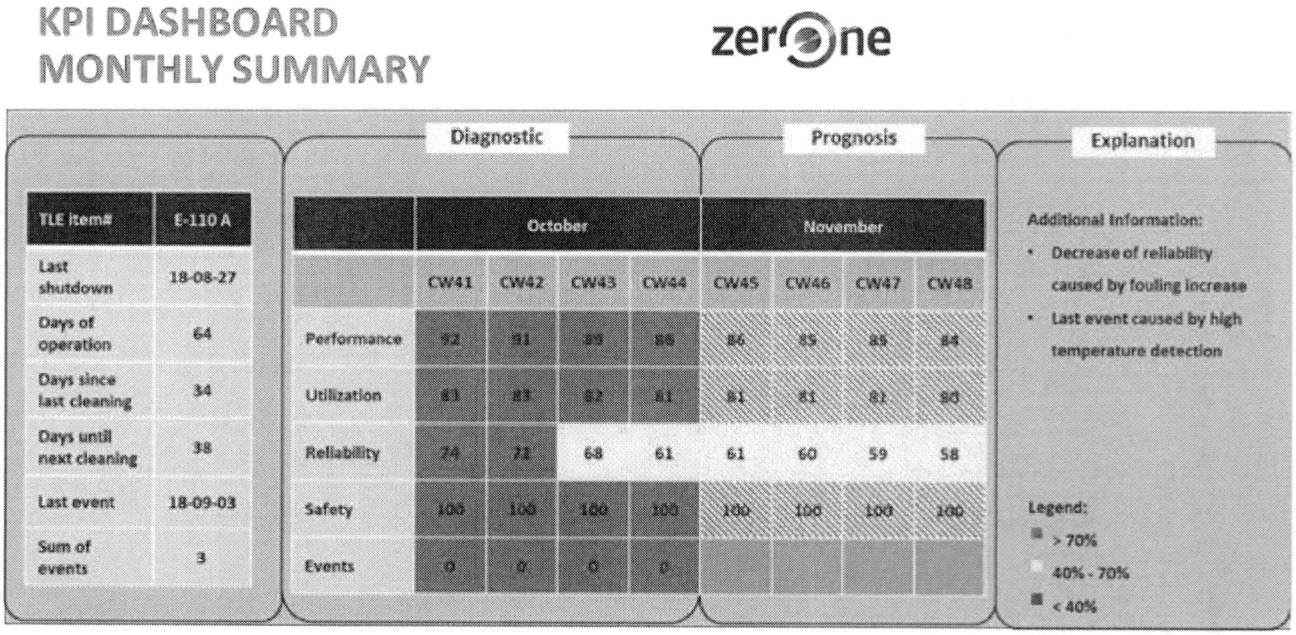

Figure 3: KPI Dashboards for single TLE

This dashboard looks clear and simple, however to derive those KPI's from sensor readings the application of OEM's design knowledge and design algorithms as well as AI is required. The user can drill down the dashboard to get more detailed information about thermal and mechanical conditions. In the following several use cases are presented to provide a deeper insight into the capability to monitor the current and future status of the TLE.

Use Case "High Inlet Temperature"

During operation the operator of a non-Ethylene plant recognized an increase of syngas inlet temperature. For a period more than two hours the temperature could not be controlled and the actual temperature reading was 90°K higher compared to the value which was used to design the heat exchanger. The operator was concerned about a potential impact on his equipment. Since Zero.One was not installed he contacted SCHMIDTSCHE SCHACK for support and to assess this event. Several discussions with the operator were conducted to clarify the situation, operational data were requested and the data were evaluated. Overall this took some days to get the right information.

In contrast to that procedure, with Zero.One the operator could have access to the web based system and could look at the condition dashboard as shown in Figure 4. This dashboard provides more details of the current TLE status. All graphs show the minimum, maximum and average value over a defined period of time as bar chart. These values are compared to upper limits like design values (shown as horizontal lines).

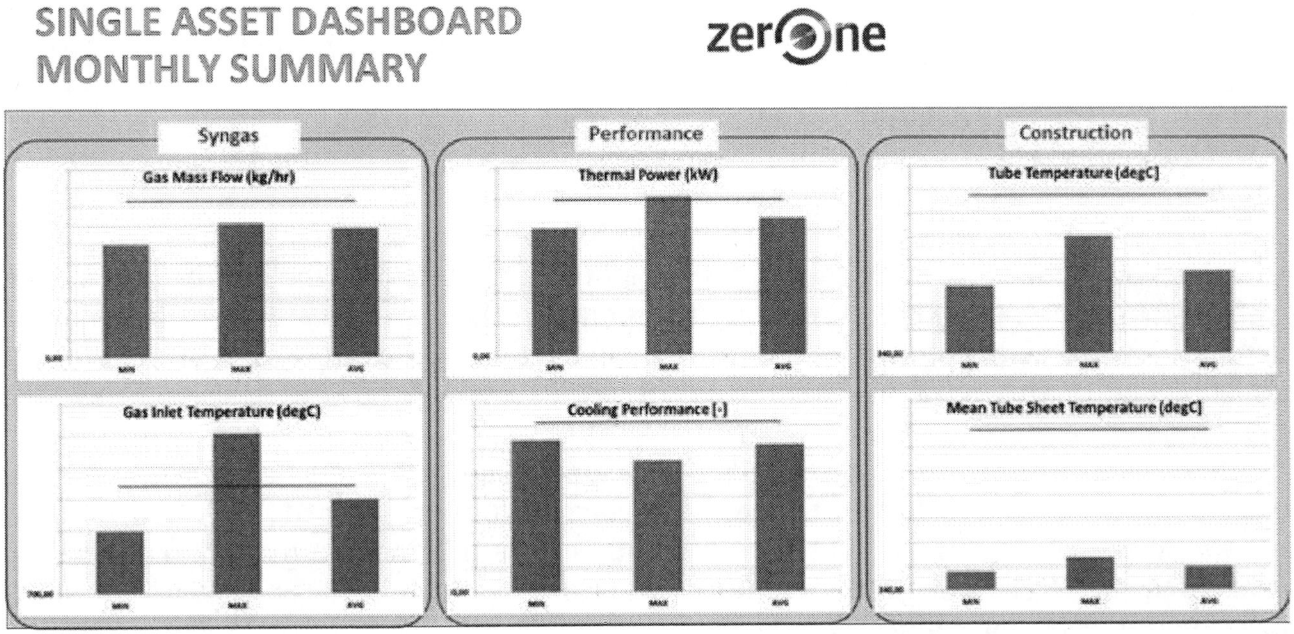

Figure 4: TLE condition dashboard for use case "High temperature event"

In the left column the values are shown which are important for the heat input into the TLE. It can clearly be seen that the syngas mass flow is always below the design value, whereas the high temperature event leads to an exceedance of the syngas temperature used for the design. Although the syngas mass flow is low, the high inlet temperature leads to a thermal power close to design value as shown in the middle column. Nevertheless the cooling capability is still excellent since all values are below the limiting value. Consequently the metal temperatures of the process tubes as well as the tube sheet are still below the design values as shown in the right column. Hence this high temperature event did not have any impact to the mechanical integrity of the equipment. This information can be obtained quickly by accessing the web based system. In addition, in case of a critical situation the operator will be informed automatically by email.

Although the information in Figure 4 looks simple, the procedure to provide thermal performance and mechanical integrity information requires the application of OEM's algorithms to calculate flow, heat transfer, hydraulic water circulation and mechanical. It is worth to mention that those algorithms are not based on commercial software but on in-house developed calculations which originally are used to design the heat exchangers. Those calculations are verified since decades by the TLE fleet operating at customers` Ethylene plants.

Use Case "Decreasing Water Side Pressure"

The second use case happened in the past, where Zero.One was also not installed. An operator informed about a damage of the exchanger and was asking for the root cause of failure. After several discussions the customer incidentally reported that after a regular maintenance shut-down the water side pressure level was decreased for further operation (the motivation to do that was not explained). A lower water pressure in steam drum and TLE results in a lower saturation temperature which indicates a better cooling capability. Although this is in general correct, it can be very critical to draw this conclusion for existing equipment which was designed for different conditions.

Assuming that Zero.One would have been installed the appropriate algorithms would have been applied based on operation data. This procedure transforms the raw sensor readings to smart data with valuable information for the operator. The results are displayed in the TLE condition dashboard as shown in Figure 5. The heat load to the TLE by the crack gas is very stable since minimum, maximum and average values are very similar. The maximum gas mass flow exceeds slightly the design value whereas all other crack gas values are below (see left column). The maximum thermal power also exceeds slightly the design value whereas the average value is also below. However, these KPI's are no cause for concern. In contrast to that the cooling capability always exceeds the limiting value (see middle column). This already indicates that the TLE cooling at water side is not further secured. The impact on the mechanical integrity is shown in the right column. The metal temperatures of the tubes as well as the tube sheet are always above the design value which directly indicates the negative impact on the mechanical integrity.

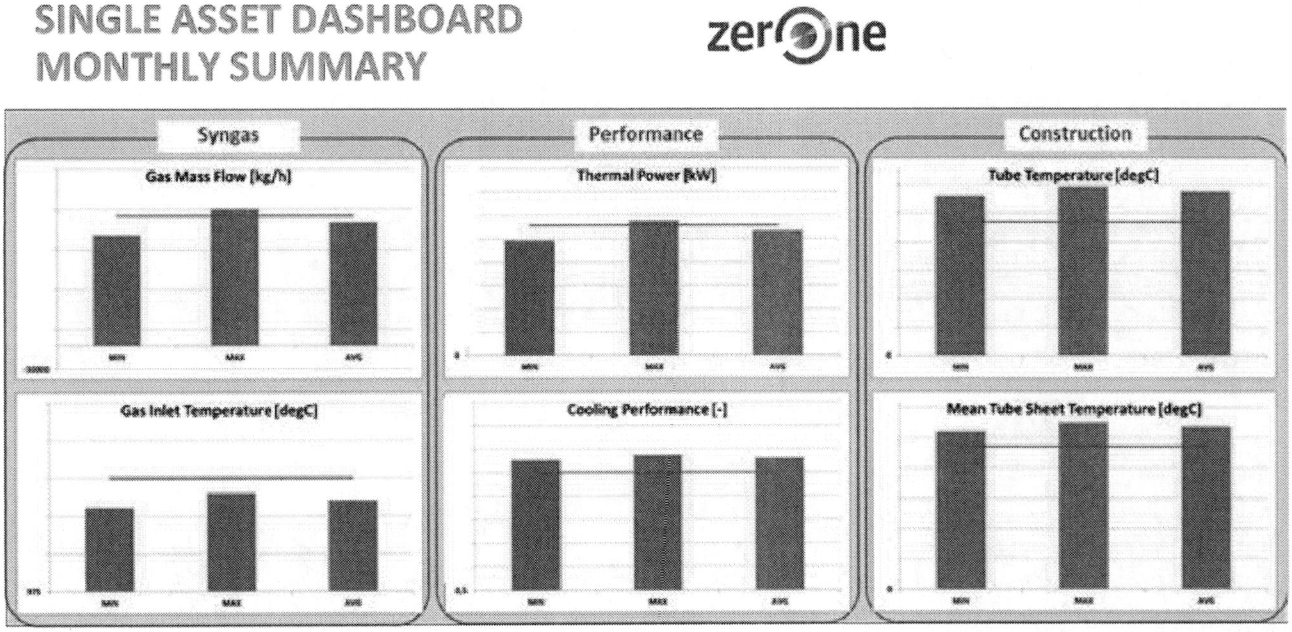

Figure 5: TLE condition dashboard for use case "Decreasing water side pressure"

To identify the root cause of the reduced cooling capability the operator can drill down to the cooling performance dashboard as shown in Figure 6. On the right side the measured sensor data for water/steam pressure and temperature are displayed as minimum, maximum and average value. The black bars describe the situation before the pressure level at water/steam side was reduced because those values are close to the design values (marked by the horizontal line). The status after reducing the pressure level is shown by blue bars. On the left the cooling performance map of the TLE is shown. The black dots that correspond to the black bars on the right are time dependent operation points. All points are placed along the expected cooling operating line that is located in the area for secured cooling capability. In contrast to that the blue dots are located in an area marked by red color for which proper cooling is not ensured. The reason for this is the decrease of water/steam pressure as shown at the right side. With this dashboard the operator directly can see the effect of lowering the pressure level. An active Zero.One system would have sent an alert message to the operator when the actual cooling performance would have reached the unsecured area. With that information the operator has the chance to apply countermeasures that bring back the performance to the valid operation range. In consequence the risk of equipment damage is significantly reduced.

The physical explanation of that use case is that due to the pressure decrease and the lower saturation temperature at water side with nearly unchanged heat load from gas side consequently the heat flux increases. In parallel, the critical heat flux as threshold for departure from nucleate boiling decreases as well as the steam fraction increases. This leads to the situation of local film boiling that dramatically reduces the cooling effect. In consequence the material temperatures increase. All calculations which describe this behavior and assess the current operation are included in the Zero.One algorithms to create the cooling performance map.

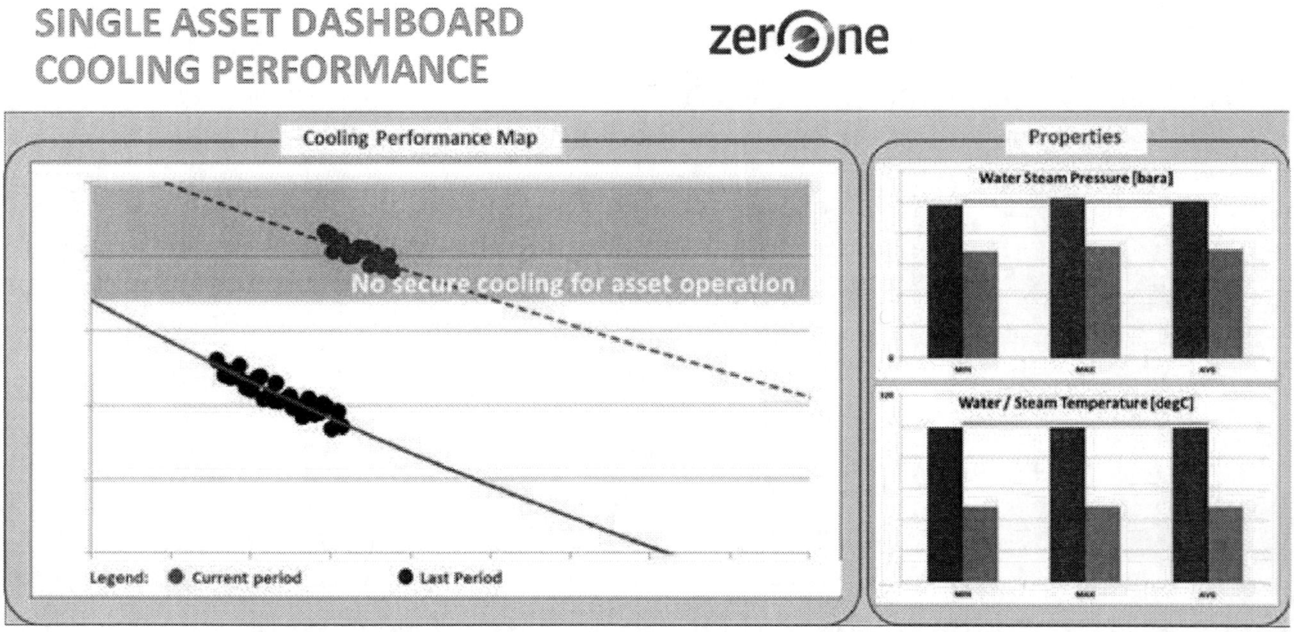

Figure 6: TLE Cooling Performance dashboard

Machine Learning

The KPI dashboard shown in Figure 3 provides also information about event or abnormality detection, prognosis of future KIP's describing the TLE condition as well as expected time period until the next de-coking. This requires the application of statistical methods, AI or the combination of AI with physical models, depending on the application.

The challenge when applying AI is to identify the right approach and algorithm. After this is obtained the model is going to be trained in a way that the error between output from the model and the known target value is minimal. When training a model, further verification is of the trained model is required to verify

- how the trained model reacts to new and unknown data
- if the training did not specialize the model in that way that accurate predictions are only possible with the training data

One use case for machine learning is to predict the time period until next de-coking of the TLE. The signal of gas outlet temperature was used as training input. The TLE is operated with different feeds, first with heavy feed then the feed is changed to light feed until de-coking. After that the operation sequence again starts with heavy feed and again switches to light feed.

The training was performed only for three operation cycles as mentioned above. After the training the algorithm was applied to unknown test data to predict the remaining time until de-coking. In addition, several unknown boundary conditions were implemented (see Figure 7):

1. heating surface reduced by 15%, 100% load, reduced operation time
   (blockage of tubes reduces heating surface, for 100% load the gas outlet temperature increases faster, in consequence operation time is reduced)

2. heating surface reduced by 15%, 90% load, normal operation time (operator reduces load to 90% to achieve the normal operation time)
3. no reduction of heating surface , 90% load, extended operation time (operator reduces load to extend operation time)
4. no reduction of heating surface , 100% load, normal operation time

The algorithm predicts the time until de-coking with an average error of +2/-4 days for the unknown boundary conditions of reduced heating surface, reduced load, extended or reduced operation time. In case that the training would consist also these events, a further accuracy improvement for the prediction is expected. However, this use case clearly shows that the chosen model is feasible for this prediction.

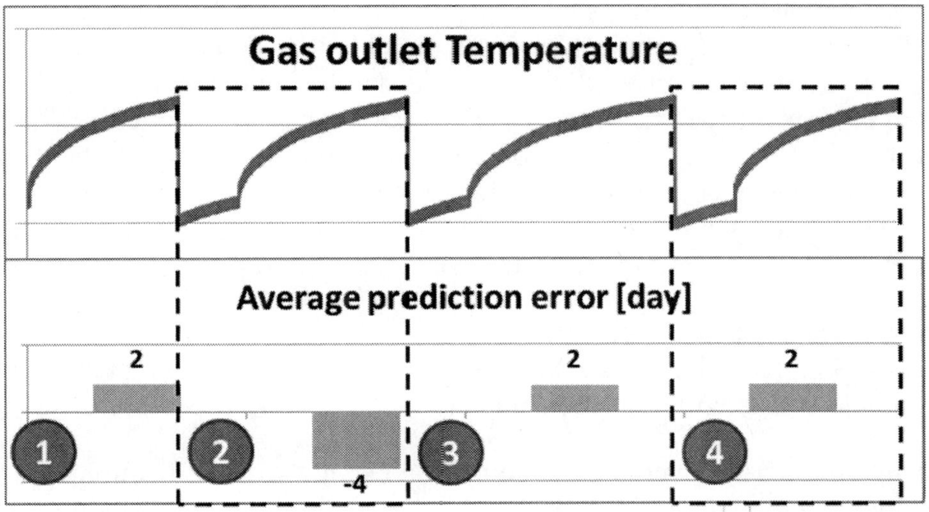

Figure 7: Test data, error between time until de-coking and prediction

Conclusion

This paper illustrates the next consequent step in further improving the functionality of SCHMIDTSCHE® Transfer Line Exchanger by applying new technologies. Zero.One comprises additional sensors installed at the equipment and applying a cloud based platform to analyze and assess the TLE. This service combines OEM design knowledge and artificial intelligence to describe the current status of the TLE, to detect abnormalities during operation and to predict future events. This creates transparency in the TLE condition, helps to optimize availability and productivity and improves planning reliability for the operator and the maintenance team.

The next possible integration steps are:

- online erosion / corrosion damage detection
- coil coking behavior for cleaning prediction
- further mechanical models for service assessments.

551675 Effective Application of Data Analytics for Ethylene Furnaces

Data analytics has become a standard tool for reliability improvement. With sufficient data and failure events, failure signatures can be recognized much earlier than might otherwise be possible. This approach can be improved upon by principle component analysis. Multivariate analysis allows recognition of a change in variable relationships. Triggering warnings when variable relationships are not following historical convention allows an even earlier indication of failure potential versus following a single variable approach.

However, an indication that relationships between variables has changed must be understood before action can take place. The goal is to go from just having a possible indication that something has changed to being able to have predictive and prescriptive advice. One successful approach to doing this is to bring expected performance from equipment specific sources such as pump or compressor curves and domain expertise to the problem and build a model of the expected performance depending on the status of each variable. An example will be shown where an incorrect installation of the bearings on the cracked gas compressor was resulting in high bearing temperatures and variable constraints and options to keep the bearing temperature down until the next shut down were modeled to allow continued operation.

Incrementally applying data analytics to Ethylene furnace to optimize performance can similarly result in very effective tools. Simple analytics looking at feed conditions and historical best can provide insightful guidance. An example of this will be demonstrated.

Taking this concept to the next level is possible by establishing variable relationships where machine learning can be applied on both historical data and automated first principle model runs to accurately predict runlengths (i.e. coking/ CPRs / TMTs) based on feed and cracking conditions. Getting value from this requires being able to understand the implications of adjustments and being able to push furnaces to match optimized performance while staying within a decoke schedule. A demonstration of how this modeling results in a "digital twin" enabling predictive and prescriptive advice will be shown.

Data analytics greatly increases the power of fundamental models and domain expertise by enabling real time feedback to operators and engineers. When applied to Ethylene furnaces, significantly more production is possible without risking excessive coil life diminishment.

550088 Using Machine Learning to Develop Deeper Understanding of Agitated Filter Dryer

Drying is a key unit operation in the manufacture of active pharmaceutical ingredients (API). A contained environment is necessary for potent APIs, which demands the use of agitated filter dryers (AFD) for filtration and drying. Understanding of the challenges and mechanism of drying is extremely important to ensure manufacturing consistency, yet despite being an essential downstream process in API manufacturing, it is one of the least understood unit operations. Challenges such as agglomeration and attrition pose threats towards attaining the desired physical properties of particles. Incomplete understanding and inefficient drying protocols can significantly affect project timelines and increase the waste of highly costly APIs.
Many experimental and computational methods have been developed to investigate the reason(s) behind all the bottlenecks, but they are still inadequate due to various factors such as being unable to deconvolute the complex interplay between the solvent and the powder, process stochasticisity, scale-dependency at manufacturing level, and computational time and resources. In this study, we have elegantly combined machine learning with discrete element method (DEM) and experimental approaches to find innovative solutions and ways to overcome the aforementioned challenges in order to design a scale-up protocol for the agitated filter dryer. Some of the key highlights of this integrated approach are prediction of the mixing time required to reach homogeneity, agglomeration probability, and attrition probability.
Similar methodology can be used to study other complex processes struggling with poor understanding of mixing, agglomeration, and/or attrition.

551237 Efficient Use of Computational Model Data Via "Zonal" Modeling

Computational models have become an integral part of the modern chemical engineer's toolkit. Such models are very powerful in representing and simulating the fundamental physics of a given process. For instance, computational fluid dynamics (CFD) and discrete element method (DEM) models are used extensively in studies relating to flow and mixing behaviour in fluid and granular processes respectively. One of the key challenges of using CFD and DEM models is their inherent computational complexity. Typical simulation run times for these models range from a few hours to a few days (esp. for DEM models). Due to this reason, such models tend to not be available for widespread consumption in the larger process engineering community.

We present a methodology here, variously called "network-of-zones", "compartment-based" and "multi-block" modeling that enables a reduction in the complexity associated with such models. Zonal modeling is essentially a spatial-averaging of simulation data that allows for information to be represented at just the right level of granularity. We will describe this approach and its relative merits as compared to traditional dimensionality-reduction techniques such as PCA. We will present some examples of application of the methodology to practical engineering problems - and also indicate extensions where it can used in conjunction with techniques such as PCA, SVM etc.

548543 Solubility in the Age of a Data Lake

Crystallization is an important process often employed several times within synthetic routes for drug substances. The development of robust crystallization processes can be separated into two tasks: (1) the selection of a solvent-system, and (2) the design of a process within that solvent system that affords materials with desired chemical/physical purity and meets yield requirements. An end-to-end workflow relying on data science techniques was developed for capturing, visualizing, and interpreting solubility data to facilitate the consistent and rapid execution of these tasks.

This workflow starts with the collection of solubility data using standardized equipment sets and approaches into templated tables. These tables contextualize the solubility data by joining each measured value (concentration, temperature, composition, etc.) with relevant meta-data (solute purity, x-ray diffraction results, equipment, date, etc.). The contextualized solubility data is ingested within a database – providing a single source for all solubility data. A templated visualization then consumes the data from this source. Further it can be filtered within the visualization as necessary (e.g., limited to a specific solute and limited to only data collected for a specific lot of solute).

To facilitate solvent selection as the first task in crystallization process development, this visualization automatically applies a decision tree to collected solubility data to classify solvents with regards to crystallization as solvents that are likely: “good for a thermal process”, “solvent within antisolvent driven crystallization”, or “good antisolvents”. Based on this classification the scientist working with the system can either begin the task of process design or apply a predictive solubility model that has been integrated within the visualization to determine other solvent systems that may be worth investigating. The application of this model uses simple R scripts with open source libraries (e.g., non-linear optimization packages) doing the “heavy-lifting”.

Once a solvent system has been selected, the task of process design begins using an automated script to fit and select the best-model for solubility data across ranges of temperature/antisolvent ratios within that system. The contextualization of the solubility data allows for visual identification and rapid exclusion of outliers within the model fitting step. Once a solubility model has been selected and fit for a given system – a constrained optimization algorithm is applied to determine the process that affords the highest yield given user supplied constraints (which are modified to ensure the process meets desired chemical/physical purity needs). This initial crystallization process conditions are then attempted and refined as necessary.

This end-to-end workflow has resulted in significant time savings, and allowed for the setting of consistent expectations for initial designs across projects. Further, it is an early demonstration of the integration of modelling and data science techniques within process development.

550152 Powerful and Novel Nonlinear Modeling and Multivariate Statistical Approaches in Big Data Sets in Chemical Processes and in Data Mining with Applications to Bio-, Medical-, and Material-Informatics

Advanced statistical methodologies have key roles to contribute in modeling chemical process data, data mining and informatics for large data sets. Over the years, our research has developed a number of statistical techniques for modeling plant data, exploiting multivariate analysis and methodologies in many applications including bioinformatics, specifically, micro-array data sets in a number of applications, medical-informatics, including disease diagnosis and discovery, and in material-informatics, including the development and evaluation of material properties and testing techniques. In this talk we present the tools and methodologies that we have developed over the years and discuss their attributes and strengths. The two primary multivariate statistical methodologies that we have exploited have been principal component analysis (PCA) and cluster analysis (CA). Our approach for modeling plant data has particular strengths when inputs are highly correlated and low signal-to-noise ratio. This talk will break these techniques down for the non-expert and then demonstrate their strengths in handling large data sets to extract critical information that can be exploited in modeling, control, analysis, inference, diagnosis, and discovery.

552246 From Data to Knowlege: AI Applications to Chemical Engineering

With the advent of the information age almost every area of human civilization has been touched in some way by the vast computing power that is commonly available. A spillover of the increasing digitization of human activity is the creation of enormous databases and online logging of exabytes of data. Outside of this titanic buildup of data is a vast gap between data storage capabilities and people's ability to understand and interpret patterns from such large amounts of data. Chemical engineers have not ignored this trend of digitization. For decades, engineers have used data from chemical plants to monitor the state of processes and maintain safe and efficient operation. Nearly all modern processes operate in a highly integrated and automated environment with a distributed control system (DCS) to maintain a consistent product quality and continuously adjust for disturbances entering in the process from the surrounding environment, with the human operator acting as a supervisor.

Data mining and knowledge discovery techniques drawn from computer science literature can help engineers find fault states in historical databases and group them together with little detailed knowledge of the process. This presentation will covers some of the activities under investigation at the Laboratory for Process Systems Engineering at Department of Chemical Engineering at LSU using machine learning tools. Focus will be on recent developments in areas such as: advanced data analytics, intelligent/smart process monitoring and development of novel image-based sensor technologies. Industrial applications will be provided.

551857 Business Value, Work Process, Analytics and Data in the Chemical Industry

This presentation will begin with a discussion of the major categories of business value that can be unlocked by applying analytics to production activities, including asset reliability, production optimization, energy efficiency optimization and safety performance improvement. It will then explore how value is realized through transformation of work processes to identify several key requirements of digitalization programs and analytics, suggesting answers questions such as:

- Should I use the cloud for analytics?
- Is my process and predictive technology data sufficient to support analytics?
- Do I need to use artificial intelligence?
- How important are metadata (metadata are plural!)?

The presentation will include a demonstration of first-principles-based analytics approach applied to early detection of asset failure in support of a top quartile maintenance work management process.

551812 Prognostic and Health Management Using Digital Twin

Digital Twin is a digital representation of your real-world device or operation that mirrors actual equipment and operating conditions. Leveraging this technology for your equipment's predictive health and operational optimization can significantly improve your company's EBITDA.

However, creating digital twins for your assets continues to be a costly affair. A typical refinery or plant has hundreds of assets. Modeling them in a robust manner is a herculean task. We will present a methodology driven by AI that works with variations in model types within a class of assets but also across different types of assets. As a result, you can rapidly scale analytics across your organization delivering extraordinary savings to your company's bottom line. Our patent pending Asset Agnostic Prognostic Technology has shown the efficiency of this approach across a variety of mechanical equipment (pumps, compressors and motor bearings. DeepIQ's prognostic health management software is

1. *a) Asset Agnostic: is easily configurable for new assets*
2. *b) Source Agnostic: connects to a wide variety of time series industrial sources*
3. *c) Platform Agnostic: works natively on both on-premise or Cloud big data environments.*

551047 Microsoft - Empowering the Chemical Industry

Chemical companies are revisiting their own business processes and technology approaches as competitors and partners start to employ ‘digitalized’ business processes and exploit the increasing convergence between OT and IT on the plant floor, to connect the enterprise as a whole to the extended supply chain and throughout the ecosystem.

In this session you will learn how technology empowers workforce with devices, while AI-powered productivity tools enable them to work more effectively, manage manufacturing complexity, and collaborate specifically to their business and operational needs. Understand Microsoft approach to Firstline Worker including Mixed Reality with introduction to Microsoft Remote Assist and Layout.

550764 Text Mining and Analytics

Text mining seeks to extract information and insights from textual data. In particular the distillation and summary of documents is of increasing commercial interest, from news media to patents to maintenance logs.

Application areas include business analysis, scientific discovery, predictive maintenance, marketing and communications, litigation, ad cybersecurity, among others.

JBasic workflow elements will be reviewed in this talk, including: definition of the corpus (which may involve the pre-step of converting images to text), preprocessing, visualization, filtering, statistical analysis, work embeddings, and classification.

Some more advanced subjects, for example topic modeling and sentiment analysis, will be introduced and described.

Demonstrations of functionality will be presented using MATLAB.

550044 Text Analytics Applications for Material Manufacturing

Global manufacturing companies generate vast amounts of unstructured data from within their operations and enterprise resource planning (ERP) systems. These text-based data sources can be difficult to integrate into traditional reporting and analysis workflows that require querying and numerical calculations. With the advancement and ease of entry into developing text-based models and features, Dow is employing several types of text modeling methods to improve margins, reliability, safety, and our customer's experience.

Dow has developed sustainable event recording processes that provide free-form text fields to precisely record the events and actions taken, but are difficult to then aggregate into reports or understand trends. In this talk, we will share examples of how Dow has developed machine learning models to categorize text for efficient trend analysis, as well use deep learning to relate documents to one another. We will then show how leveraging these text analytics methods can deliver value through new insights and streamlined work processes.

551060 Opportunities for Text Mining in Service of Chemical Engineering

As the world has become increasingly digital, vast storehouses of data in digital form have become available both within the public sphere and within companies as proprietary resources. Over the past decades, there has been a growing awareness of the strategic value of such data to companies and to society more broadly in virtually every field - Chemical Engineering is no exception. The social web can be mined to monitor public opinion related to products and policies and how these change over time in response to news events. Documentation about work practices and events within companies can also be mined to identify ways in which company practices affect its outcomes, employee retention, and even safety concerns. Scientific literature can be mined to identify trends in research topics, common or competing findings, and emerging research interests. The news can be monitored to identify connections between environmental policies, company practices, and environmental impact.

Text mining is a field that brings together an understanding of natural language as an unstructured form of data and machine learning as a suite of modeling tools to layer structure over it. Using machine learning paradigms like probabilistic graphical models and deep learning, it is possible to architect modeling tools that are able to extract latent structure found in text. By identifying and extracting the latent structure found in text, it is possible to transform this unstructured data into structured data, at which point it is possible to apply a plethora of modeling tools to identify trends over time and even plausibly causal relations between events and states. A major goal of research in the area of text mining is to achieve robustness in the face of noisy data. With the big data revolution, new storehouses of vast amounts of textual have become available. While this data is far less well-structured than forms of language that were the target of work in natural language processing decades ago, now research on text mining applied to even the noisiest of this data (for example, from sources such as Twitter or Reddit) are commonplace.

This talk will offer a brief overview of state-of-the-art methodologies for applied machine learning, with pointers to resources and further instruction.

Safe and Optimized Operations Through Smart Analytics & Big Data

Karl McManus
Chartered Chemical Engineer, IChemE
Wood, Galway Technology Park, Parkmore, Galway, Ireland
Karl.mcmanus@woodplc.com

Contributions by

Jason Payne
Analytics Manager, Digital Solutions
Wood, Galway Technology Park, Parkmore, Galway, Ireland
Jason.payne@woodplc.com

Ronan O'Malley
PhD, Senior Data Scientist,
Wood, Galway Technology Park, Parkmore, Galway, Ireland
ronan.omalley@woodplc.com

Prepared for Presentation at
American Institute of Chemical Engineers
2019 Spring Meeting and 15th Global Congress on Process Safety
New Orleans, LA
March 31 – April 3, 2019

Safe and Optimized Operations through Smart Analytics & Big Data

Keywords: Predictive Fault detection, asset integrity assurance, artificial intelligence, machine learning, sensor data

Abstract

Are offshore oil and gas operating companies using all their available data to demonstrate safe operations and optimise performance?

Wood have developed applications in the areas of Prognostics (Predictive Fault detection) and asset integrity assurance. Using supervised and unsupervised Artificial Intelligence (AI) learning techniques, automated pattern detection and classification, our engineering teams have reduced risk, proven process safety assurance and reduced OPEX costs by millions of dollars per year.

The following application examples (case studies) have been developed by operational chemical process engineers working closely with data analysts. The applications manipulate big data to support client operations and save OPEX costs.

- Prognostics Case Study: Valve Actuator Performance – The Wood valve actuator performance application uses sensor data and machine-based learning techniques to help predict failures of production critical subsea valve actuators.
- Asset Integrity Case Study: Water Ingress – Traditionally, major oil and gas companies spend significant amounts of revenue on Subsea Under Water Inspections (UWI). The Wood water ingress application utilises data driven condition monitoring to demonstrate asset integrity thus reducing the requirement for UWI and minimising OPEX costs.

1 Introduction

Digitization is not new, but it is transforming the oil and gas industry. Exploration and Production (E&P) companies are utilizing technology that captures and stores large amounts of data points from wellhead sensors, valves, pipelines and various equipment installed in the field to remote monitoring stations. With this data, operational processes can become much more efficient and safer. Digitization has allowed E&P companies to use technology, big data, and predictive analytics to extract resources more efficiently from each well, create a safer work environment, and do all of this with better labor utilization.

This paper outlines two case studies to provide Prognostics (Predictive Fault detection) and asset integrity assurance. The valve actuator performance application (case study 1) uses sensor data and machine-based learning techniques to help predict failures of production critical subsea valve actuators. Assurance to operations that no water ingress has occurred in deepwater facilities (case study 2) is achieved by event detection and event classification based on modelled and real data.

2 Case Study 1: Prognostics; Valve Actuator Performance

2.1 Problem definition

After approximately three years of production a number of subsea valve actuators started to fail unexpectedly. The root cause of the failure was thought to be hydrogen induced stress cracking of the actuator springs. These failures were undetected until the valve was hydraulically actuated from the control room. Spontaneous failures result in significant downtime (months), deferred production (e.g. 10mbd well shut-in), substantial engineering and intervention costs.

2.2 Solution

Condition monitoring of actuator performance became a high-profile project for an asset due to unrevealed valve failures. With hundreds of this type of actuator in the field, this became a large data capture, review and analysis project. Subsea gate valve signatures graphically illustrate changing hydraulic pressure and flow parameters at various points in the local valve actuator circuit during the stroking operation. By identifying the characteristics and data patterns of healthy valve signatures degrading trends can be identified. Six health indicators have been identified for a valve close signature and these are illustrated in Figure 1. Over time, specific rates of change can provide useful information on how a valve is performing or degrading. The six indicators are as follows:

1. Accumulated Volume – Has the piston performed a full stroke?
2. Valve Closing Time – Are there any signs of reduced spring force?
3. Clear Stop – Is there a clear stop point indicating that the actuator has reached the end of its stroke and stopped?
4. Initial Function Pressure Drop – Indication of initial spring movement?
5. Function Line Pressure Curvature – A damaged, cracked or plastically strained spring would show a change in the force extension characteristics.

6. Pressure Differential – An excessive DP between start to fully closed function pressure will indicate a reduction in spring force.

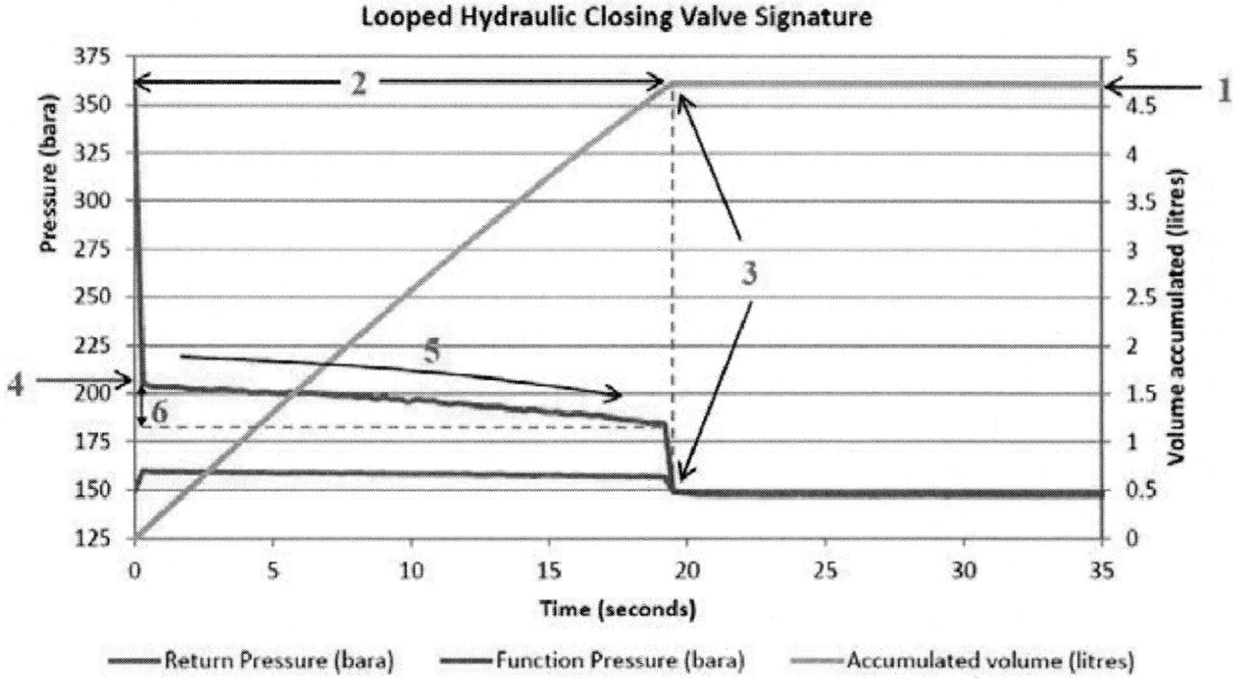

Figure 1. Healthy Closing Actuator

Figure 2 provides an example of one actuator with two signatures overlapped. The healthy signature is shown in solid lines. The same actuator in a degraded state is shown with dashed lines. By identifying early signs of degradation, the asset team can predict when a failure is likely to happen and put an intervention plan in place accordingly to minimise downtime and deferrals. An unexpected actuator failure can result in an oil well being shut-in for 2-3 weeks whilst a vessel is sourced, and intervention preparations (actuator lock out tool) are completed. This has resulted (depending on deferred cost rates) of > 1million USD lost/deferred production based on a well that produces 10mbd. By completing this activity during planned maintenance there are minimal deferral impacts.

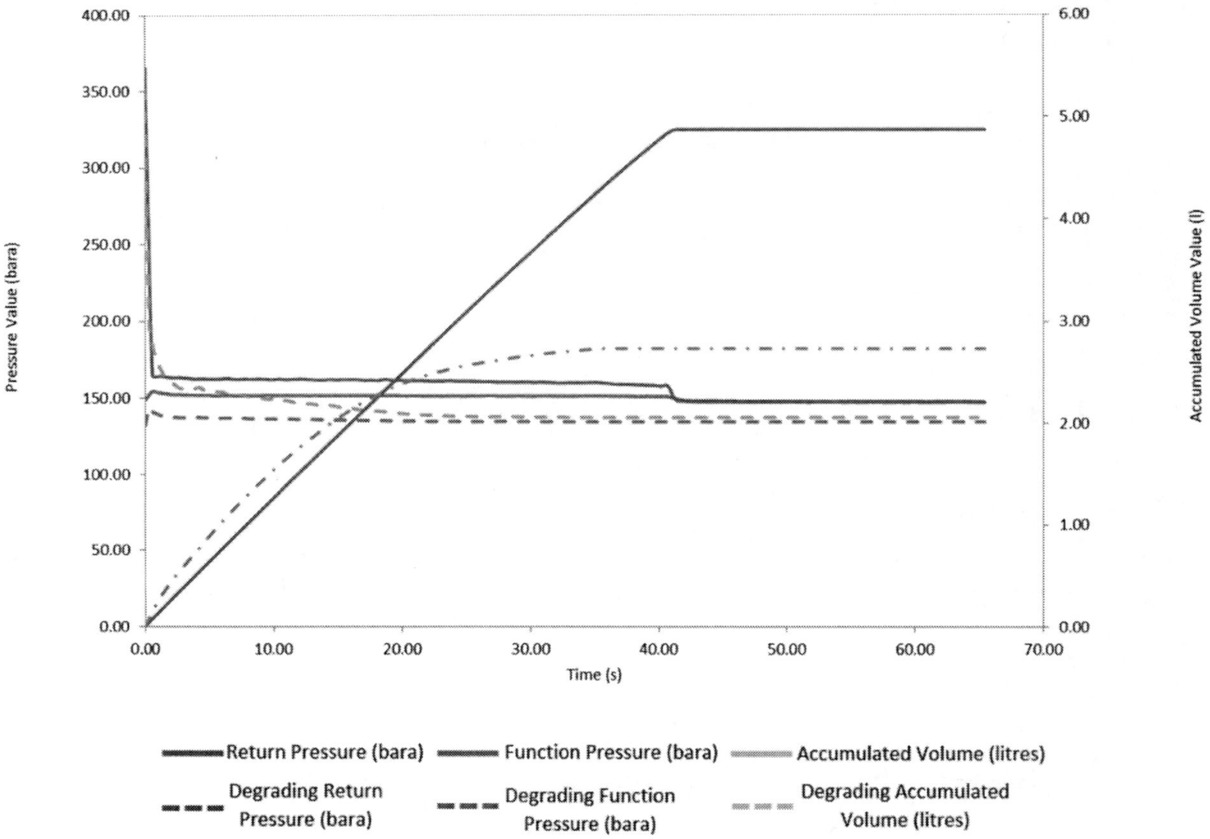

Figure 2. Example of Degradation from Healthy Signature

Currently there are more than 50,000 actuator signatures stored in the asset database. Two machine learning techniques were deployed to automate detection of well understood degradation patterns and track condition of equipment that has no recorded history of failure (or degradation):

- Supervised learning (similarity and information-based approaches) for well understood patterns
- Unsupervised learning (dimensional reduction, clustering) for equipment that has no recorded history of failure or degradation

Both techniques are deployed within human-in-the-loop applications. A human-in-the-loop application leverages machine learning or artificial intelligence to improve human performance, i.e. the human and machine work together to solve problems that can't be solved by human or machine working alone. For valve actuators, the digital process enables review of actuators at scale and at times that would be infeasible without the digital process.

Supervised learning currently employs a k-Nearest Neighbor (kNN) approach (where there are known and well-understood degradation features) and suitable for to the desired human-in-the-loop philosophy. The current algorithm has a greater than 90% accuracy at predicting both healthy and degrading actuator signals.

As actuators are imported from the field, the algorithm automatically performs a first pass classification of health and flags potential issues. When reviewing actuator signatures hardware engineers are presented with the machine learning classification and the basis for the classification, e.g. the three nearest neighbors from history. Nearest neighbors from history are signatures that have been classified in the past by hardware engineers and have been determined by the machine to be similar to the new signature. A majority voting system is applied by the machine to determine health classification, e.g. if 2 out of 3 neighbors are healthy then the signature is classified as healthy. Cases where the machine doesn't find similar neighbours requires engineering intervention. This manual classification is used by the machine in future assignments.

For assets with no recorded history of failure or degradation, unsupervised learning is employed to monitor the condition of actuators based on classification and association with existing human-labelled clusters. The algorithm forms a cluster, the hardware engineer examines the cluster and tags it based on domain expertise, e.g. healthy sensor with data errors or healthy operating. If a new cluster is developed or is developing this is automatically flagged by the machine and brought to the attention of the hardware engineer. This new or emerging cluster can then be efficiently reviewed within the application and tagged by the engineer, e.g., cluster to watch or potential degrading cluster formation. Each time a new signature is imported by the application it is plotted and the strength of its association to other known clusters is calculated.

By training and deploying machine learning models (supervised and unsupervised) to automatically identify these patterns results in giving approximately 18 months' notice before 'failure'. This is a significant step towards digitalization of workflow, significant reduction in manhours and improved inspection intervals and service to client. Digitalization makes it feasible for engineers to monitor, leverage and report on valve actuator data at regular (normally monthly) intervals. With a cloud hosted web application (workflow interface for engineer) this enables engineers based in all parts of the world access to database technology and connectors to historian/source data as required.

3 Case Study 2: Asset Integrity; Water Ingress Detection

3.1 Problem Definition:

Offshore oil and gas companies typically spend millions of USD per annum on inspections. A big risk to deepwater subsea assets is water ingress. At typical operating pressures (<100 bara) downstream of the choke valve, for deepwater oil-producing assets, a breach in equipment or pipework will result in water ingress due to relatively high hydrostatic pressure (~>130 bara). Water ingress has the potential to cause hydrate formation in the pipe which can lead to blockage. Inspections are performed at routine intervals and cannot detect water ingress.

3.2 Solution

The intent of this project was to use data analytics methodologies to strengthen data driven condition monitoring and reduce inspections and minimize OPEX of existing assets. Artificial intelligence can recognize patterns (neural networks) and identify healthy or abnormal behaviour operating conditions such as sea water ingress into the system. For Artificial Intelligence to recognize an ingress of water, it was fed pressure/temperature data of what this scenario may look like based on flow assurance modeling. A flow assurance model was created in OLGA to represent the ingress of sea water in the system. The intent was to pilot an approach of utilizing data generated from transient models developed and use this model to teach the machine what a water ingress event looked like. The model starts from the reservoir and ends at the manifold exit. Figure 3 outlines the leak location modelled on a manifold.

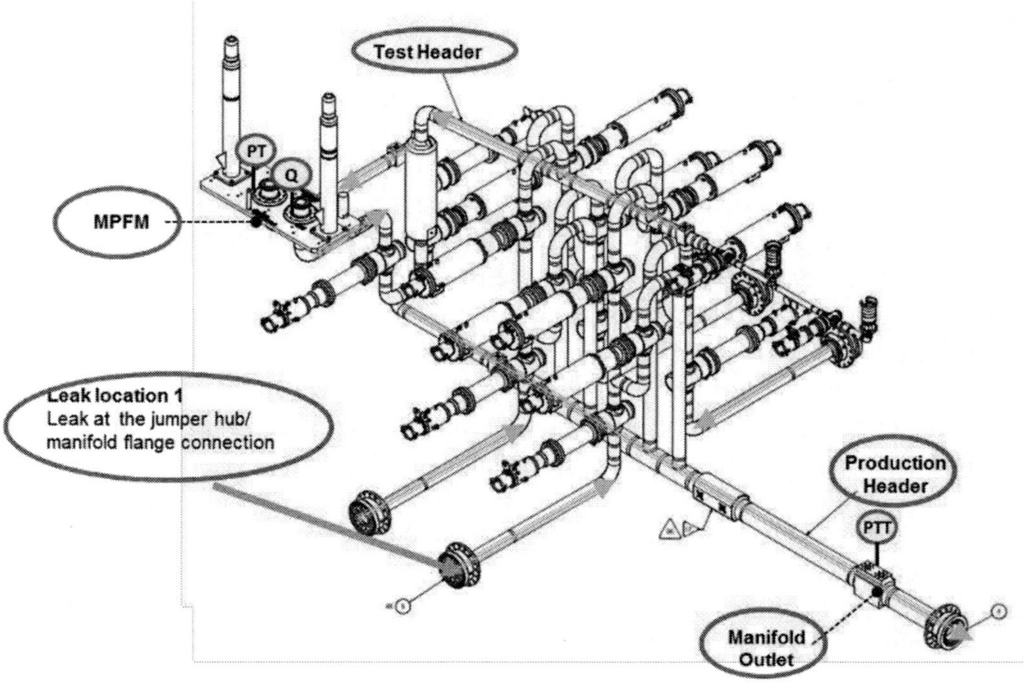

Figure 3. Manifold Isometric Pipework

Changes in temperature, pressure, choke position, liquid flow rate, liquid velocity and water-cut over time at the manifold exit due to sea water ingress were investigated. Using the modelling results software agents were developed to monitor the relationship between temperature upstream of the choke in each connected tree and production header temperature at the manifold exit. To initiate a temperature agent for a location, event detection and event classification takes place. The previous two years of real data from relevant instrumentation was queried from the historian at thirty-minute resolution. A linear model was fit between the independent variables (tree temperatures) and the dependent variable (manifold temperature). While the linear model was not an optimal representation of steady state operations, deviations from the model are a reasonable mix of ramp up / ramp down of production, explainable temperature anomalies (mainly due to sensor malfunction) or false alarms.

While detecting the water ingress simulations provided from the flow assurance model, a new variable called the harmonized signal is derived from the linear model. This harmonized variable is derived from the absolute value of the difference between successive residuals.

$$\text{harmonized} = \text{abs}(\text{diff}(\text{residuals}))$$

To extract events from this signal, a threshold is derived from its distribution.

threshold = mean(harmonized) + 3*sd(harmonized))

Any connected group of points was converted to an event at this location with an associated start and end time. This even boundary was widened by five hours each side and combined with any other events that result from the widening. As the event detection process was designed to be conservative, a second layer of classification was applied to label the event as normal or abnormal. This considers other instrument data around this location during the timeframe of the event to decide which label to apply. After experimentation it was found that the choke position data provided the most relevant information for automated labelling.

$$choke\ movement\ feature = \log(\max(choke\ position) - \min(choke\ position))$$

Figure 4 shows a 2D dimensionally reduced representation of events for a single manifold with highlighted normal events and anomalies.

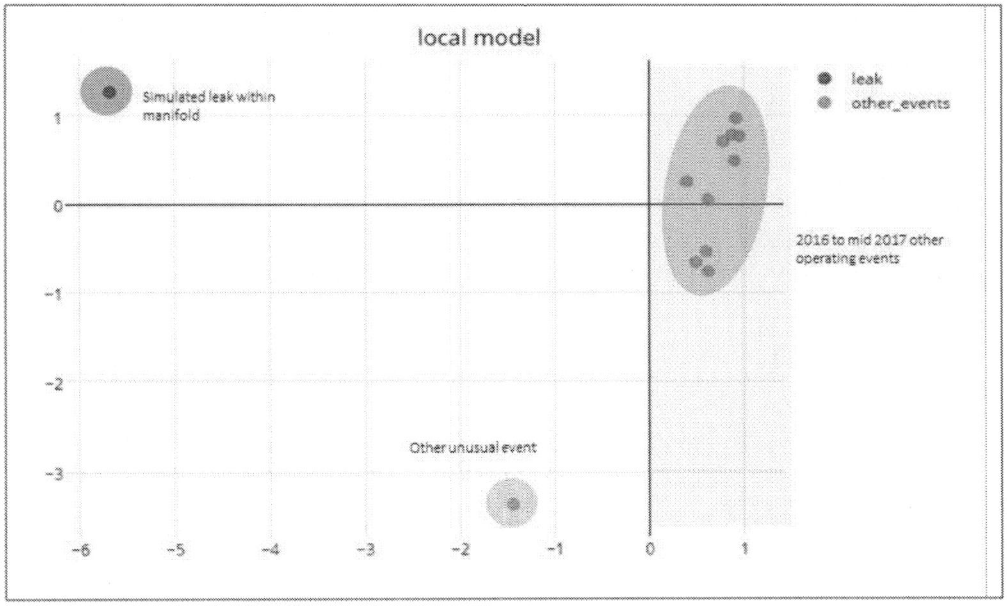

Figure 4. A 2D dimension reduction of events for a manifold.

Agent updates are performed on a regular basis ahead of scheduled engineering review, where new events are added to the cloud based web application. Event data can be analysed, reviewed and labelled by engineers during the regularly scheduled review process. By combining simulated leak data with historical field data and using advanced analytics, a strategy for deployment involved employing numerous autonomous agents each monitoring a section, rather than a single agent monitoring at a high level. Additional agents will be added to scans over time to increase sophistication.

4 Conclusion

When operating facilities and plants, engineers and managers need to be proactive in relation to process equipment performance. In the case studies discussed, with the utilization of historian data (process/hydraulic pressure, flow, temperature etc.) available it is possible to use technology, solutions and predictive analytics to make informed decisions more efficiently and create a safer work environment.

The case studies involve human-machine interaction, but the machine does not make the final decision. Competent process and operations engineers are still very much integral to the overall outcome. These case studies have provided significant positive impact on risk-based inspection (significant cost saving), data-driven innovation, automatic notification of potential anomalies or operational performance issues before they become significant. Companies can plan for the failure of equipment, minimize downtime through planned intervention/repair, reduce installation costs, have temporary/ standard operating procedures written (and approved) in advance of the change as part of the management of change process. Early identification of possible failure results in significant cost savings and ultimately minimize or eliminate production deferrals.

Data analytics is a continually evolving. To keep abreast of latest technology, the data analytics team in Wood are working on a research project (PREMSAM) with the Centre for Applied Data Analytics research (CeADAR, Ireland). These collaborations are envisaged to further develop methodologies for condition monitoring of process safety equipment performance. By developing innovative digital solutions to technical integrity management, operations and process safety management as outlined in the case studies above we hope to make industry safer, smarter and more cost effective.

5 References

[1] C. Aldrich, L. Auret, "Unsupervised Process Monitoring and Fault Diagnosis with Machine Learning Methods", Springer-Verlag, London, 2013

[2] Hastie, T., Tibshirani, R., Friedman, J. (2001). *The Elements of Statistical Learning*. New York, NY, USA: Springer New York Inc.

[3] Kelleher, J. D., MacNamee, B., D'Arcy, A. (2015). Fundamentals of Machine Learning for Predictive Data Analytics: Algorithms, Worked Examples, and Case Studies. Cambridge, MA: MIT Press. ISBN: 978-0-262-02944-5

Don't Worry!! Laboratory Safety Is Fun Do, Let's Play a Mobile Game ☺

Nishaben Desai Dholakiya
University of Pardubice, Faculty of Chemical Technology Czech Republic
E-mail: st53279@student.upce.cz
Milos Ferjencik
University of Pardubice, Faculty of Chemical Technology, Czech Republic
E-mail: Milos.Ferjencik@upce.cz
Josef Brozek
University of Pardubice, Faculty of Electrical Engineering and Informatics, Czech Republic, E-mail: Josef.Brozek@upce.cz
Zdenek Jalovy
University of Pardubice, Faculty of Chemical Technology Czech Republic
E-mail: Zdenek.Jalovy@upce.cz
Jan Kubik
University of Pardubice, Faculty of Electrical Engineering and Informatics, Czech Republic, E-mail: st52531@student.upce.cz

Prepared for Presentation at
American Institute of Chemical Engineers
2019 Spring Meeting and 15th Global Congress on Process Safety
New Orleans, LA
March 31 – April 3, 2019

Don't Worry!! Laboratory Safety Is Fun Do, Let's Play a Mobile Game ☺

Nishaben Desai Dholakiya
University of Pardubice, Faculty of Chemical Technology Czech Republic
E-mail: st53279@student.upce.cz
Milos Ferjencik
University of Pardubice, Faculty of Chemical Technology, Czech Republic
Josef Brozek
University of Pardubice, Faculty of Electrical Engineering and Informatics, Czech Republic
Zdenek Jalovy
University of Pardubice, Faculty of Chemical Technology Czech Republic
Jan Kubik
University of Pardubice, Faculty of Electrical Engineering and Informatics, Czech Republic

Keywords: Laboratory Safety, Spiral learning, Kolb's ELT,

Abstract

The overall aim of this article is to change the general perspective of laboratory safety from being boring to make it more interesting using mobile games. A type of Mobile Serious Game (MSG) is underway of designing for laboratory safety at the Institute of Energetic Materials at the Faculty of Chemical Technology, Pardubice University (CZ). This application is principally aiming a group of players who, after a period of time feel either over confidence or ignorant towards laboratory safety. It is believed that challenging nature of the game would make learning precisely interesting. Three different levels of game would be based on level of difficulty with respect to safety rules. i.e., level one is all about Personal Protective Equipment (PPE), level two would highlight one of the hazardous Nitration reactions, while level three would be designed for detailed learning of safety rules connected with the reaction. Second half of this paper would argue how constructivism and learning theories are believed to be suitable options while correlating safety with MSG. Whereas Kolb's Experiential Learning Theory (ELT) offers four stages of learning; a) concrete experience b) reflective observation c) Abstract conceptualization and d) active experimentation. Which means a learner/player would enter from any stage of learning cycle to complete learning process to gain knowledge of laboratory safety.

Introduction

A Mobile Serious Game (MSG), a combination of education and entertainment. has received significant concern in the field of education, and a mobile educational gaming community has been formed in recent years [1], Mobile applications are the biggest component of Apple services and helped that segment gross over $38billion in 2017 from Applications store. Furthermore, Applications Store has more than 65,000 educational applications, primarily intended for iPad, which cover a wide range of fields suitable for different levels and learning styles [2]. With a mobile phone in hand, you will be up-to-date on the latest news, regulations, and incidents. You can quickly communicate with anyone you need, regardless of whether you are on-site. You can even pull up any piece of information you need at any time [4]. Besides, mobile apps have proved to be valuable educational tools; laboratory instructors have been limited to using mobile devices only for virtual laboratories with simulated experiments. Which recently been changed, as researchers have developed a series of mobile applications that allow students to remotely interact with real data and equipment in real laboratories. Students reported deeper engagement levels using mobile apps and the virtual lab [5].

Chemical process safety by *OSHA* (Occupational Safety and Health Administration) *Safety Regulations, Checklists Audit Reports* [6] is a mobile application provides a compilation of laws and regulations regarding the handling of hazardous chemicals, as well as a database of checklists which can be downloaded and used for audits. Another OSHA-based app is *OSHA Safety*, which comes with all OSHA safety regulations and standards for hazards pre-downloaded so they can be accessed anywhere at any time without connection to internet or a cellular network [7]. When connected to a data source, the application can direct the user to the latest OSHA news. Both of these applications are free of charge.

Considering extensive ranges of mobile phone applications in the field of education and training, we came up with a concept LABSCAPE, a type of educational mobile game application which is underway of designing to improve laboratory safety. An overall aim of the LABSCAPE is to offer a new direction not only towards laboratory safety training; however it also aimed to make safety more interesting. It offers an online platform where a game player could potentially discuss his/her difficulties with other players from any part of the world, without associated risk. A player making mistake means losing a level rather than danger of himself/others while performing real laboratory experiments.

In order to contribute to reflect on the mobile game use in learning process for laboratory safety, This article will map out some guidelines of the pedagogical proposal using Kolb's Experiential Learning Theory (ELT). This identifies and details a complete four-stage learning cycle, showing how experiences can be translated depending on learner's choice to enter the learning cycle [8]. Whereas a spiral learning theory where a learner continuously constructing knowledge based on previous experience at the end of every cycle.

LABSCAPE for Nitration

The MSG is being designed for Nitration reaction for preparation of Adipic Acid by the masters or research postgraduate students at the University of Pardubice. Based on laboratory experiment data the Nitration reaction initiates with 10 ml of water, 20 ml of 65% nitric acid (14.3 g; 147.4 mmol) was added and the solution was warmed to 85°C. Which followed by 4 ml of Cyclohexanol (3.8 g; 38.4 mmol) gradually added such a way that temperature was maintained between 80-90°C. (Dinitrogen Tetraoxide and other nitrogen oxides are evolved during this period). After addition of Cyclohexanol, the reaction mixture was heated at 100°C for five minutes. Afterwards this mixture was cooled to 15°C, where the solid state of mixture was filtered, washed with cold water and crystallized form water. The yield of Adipic acid was measured at 2.3 g (42%).with 153-155°C of Melting Point. While the fact about hazards correlating this reaction are listed as; a) 65 % Nitric acid is corrosive to skin and eyes. The vapours can irritate respiratory system. b) Dinitrogen Tetraoxide (N_2O_4) released during reaction is toxic. It causes burns in contact with skin, eyes or of if inhaled. It also causes pulmonary edema if inhaled.

Screenshot in Figure 1 highlights a welcome screen of LABSCAPE, with list of options for game players. Game player needs to remember both categories of rules; a) game playing rules b) safety rule to clear the level and gain points. Colorful graphics and real like laboratory environment is likely to encourage players to continue playing for fun. For the major part of designing were used mainly the basic operations of 3D modeling using *Blender* software. All meshes were also made with the possible smallest amount of polygons due to high-poly models could have a negative impact on performance. While software UNITY would be used for coding part of the game and finally MATLAB software would provide some simulation part of the game. It is believed that launching LABSCAPE at an online play store would maximize possibilities to get maximum reviews from game users from around the globe which eventually would assist in data analysis for its usefulness for laboratory safety.

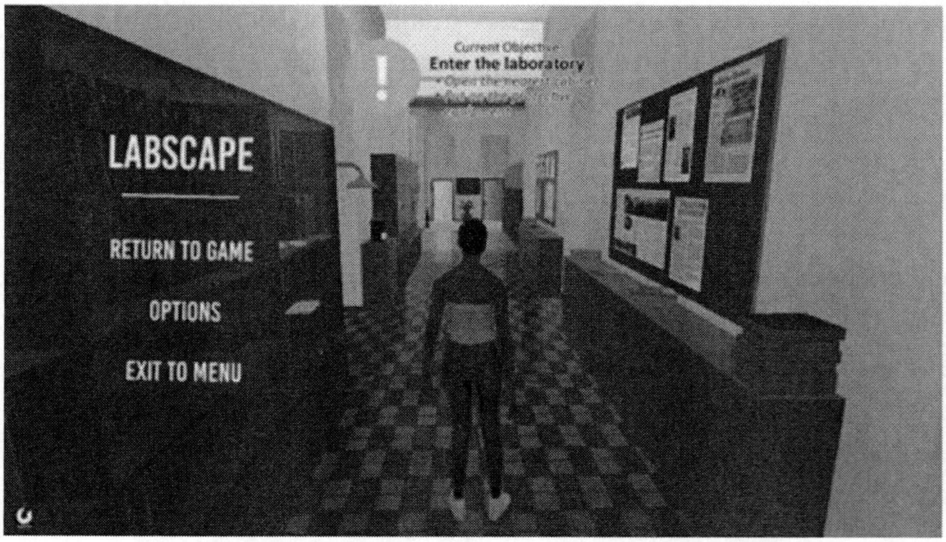

Figure 1: Screenshot of Welcome Screen (LABSCAPE)

Laboratory Safety (Level I)

While practicing engineering safety protocols in laboratory experimental procedure enhances students' commitment to improve their safety behavior and attitude toward carrying out operations in industrial process units. As students study the experimental procedure and identify hazards involved before the unit is run, potential undesirable incidents can be identified and hazard control measures can be processed within the system. This makes the laboratory environment more safe and resilient [9]. Table 1 highlights point's allocation pattern of LABSCAPE, with respect to consequences. Depending on type of mistakes and how often a player is making mistake, would decide a player to continue the same level or fail the level and start from same level. It isn't feasible to play with mobile in laboratory hours, it would be offered in free time before or after laboratory hours. It is believed that chatting with online players or coworkers relevant to LABSCAPE would make it as an open platform to correct mistakes, motivate to reentering the ELT cycle and challenge them.

Table 1: Correlating consequences

Steps	Causes	Consequences	LABSCAPE
Entering laboratory	No gloves No goggles No coat	HNO3 spill on hand HNO3 spill on body Eye irritation	Losing points/level
Pouring HNO3	Not reading label	HNO3 Spill on body HNO3 Inhale/ swallow	Losing points/level
Drinking coffee	Inside laboratory	Swallow chemical while drinking coffee	Losing level
Reentering laboratory	Without gloves	HNO3 spill on hand	Losing points/level

Figure 2: Screenshots of possible variation to enter laboratory using PPE

Figure 2 indicating the possibility of how a game player can make mistakes with PPEs before entering to laboratory. Depending on type of error, the player would lose points, while if mistakes are repetitive, he or she would likely to lose the level, which is indicated in Figure 3, with consequences of causes II and III of first step of entering laboratory. It was decided to let the player continue the level by entering laboratory even with missing PPE to help them remember the consequences of mistakes for further learning. For instant, player is dealing with HNO3 even without wearing gloves or goggles to make them analysis associated hazards as shown in Figure 3

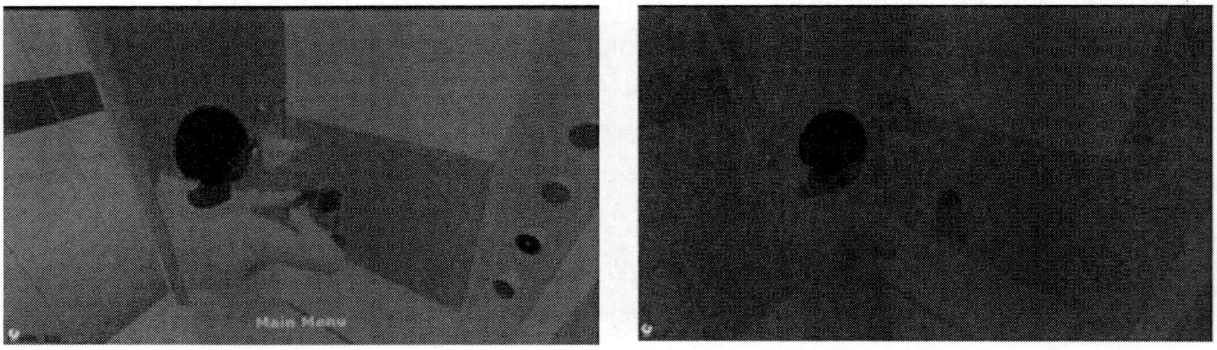

Figure 3: Screenshots of concequences of causes I and III

Kolb's Experiential Learning Theory (ELT) for LABSCAPE

Kolb defined learning as "the process whereby knowledge is created through the transformation of experience and knowledge results from the combination of grasping and transforming experience" [10]. The ELT model reveals two vertically related modes of grasping experience—Concrete Experience (CE) and Abstract Conceptualization (AC)—and two horizontally related modes of transforming experience—Reflective Observation (RO) and Active Experimentation (AE) [11]. Kolb's ELT cycle begins at any one of these four stages, but the following is one of the most often sequences suggested for the learning process [12]:

- Concrete Experience (CE): the learning process begins with the students actively carrying out an experience/activity
- Reflective Observation (RO): following the Concrete Experience stage, the students consciously reflect back on the experience;
- Abstract Conceptualization (AC): The students attempt to understand the general principles under the experience, trying to conceptualize a theory, model, or hypothesis of what is observed;
- Active Experimentation (AE): the students try to plan how to test a model, theory, and/or plan for new experiences, probably with different behaviors.

The impact and efficiency to achieve the complete navigation Kolb's learning cycle, it was believed to investigate methodological and pedagogical aspects of laboratory safety, using different levels of LABSCAPE. According to Kolb the ELT cycle is actually a learning spiral. When a concrete experience is enriched by reflection, given meaning by thinking and transformed by action the new experience created becomes richer, broader and deeper. Further iterations of the cycle continue the exploration and transfer to experiences in other contexts. John Dewey was the first to describe how the learning spiral leads to development; transforming the impulses, feelings and desires of concrete experience into higher order purposeful action [13]. A spiral ELT was introduced, where each cycle would represent each level of LABSCAPE. For instant, LABSCAPE level I defines four stages (CE, RO, AC and AE) of Kolb's ELT to learn introduction part of laboratory safety using PPE. While level two examine hazardous properties in preparation of Adipic Acid using Nitration reaction as shown in Figure 4. Level three is designed for student to understand how the changes in parameters influence consequences of reaction into which he was introduced during the level two. And finally, level four would actually be the implementation cycle, where learner would enter to laboratory to perform real experiments based on what he or she has learned in cycle one to three. Based on its application for laboratory experiment, it named as Introduce, understand, experiment and implement respectively as indicated in Figure 4.

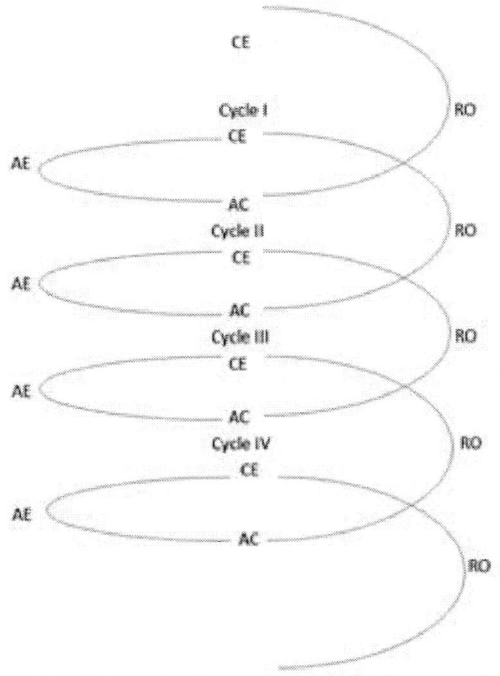

Figure 4: Spiral Learning using LABSCAPE

Table 2: Correlation of LABSCAPE with spiral ELT

Spiral	Concrete Experience	Reflective Observation	Abstract Conceptualization	Active Experimentation
Cycle I	Existing knowledge	Chatting	Developing	Playing
Cycle II	Reading manual	Brainstorming	Updating	Playing
Cycle III	Base of level II	Questioning	Developing	Playing
Cycle IV	Base of level I to III	Correlating	Correcting	Performing

Cycle I: It is designed to highlight the importance of PPE; CE would serve the gateway to enter the learning cycle using their existing knowledge. Discussing doubts or points with fellow players or chatting online with RO approach, would result in developing own concept for PPE as AC stage. This cycle would end with playing LABSCAPE (AE) one successfully, or losing the level would redirect to same level from scratch as indicated in Table 2.

Cycle II: This cycle highlights a Nitration reaction to prepare Adipic Acid which is detailed earlier in this article, would serve a CE stage, where a game player would recollect his experience for the reaction. Table 2 shows how Brainstorming season required to update existing knowledge of the reaction would offer RO with outcomes the cycle by constructing own ideas in AC stage to result in playing the game for AE.

Cycle III: Experience from cycle II would work as base knowledge to enter this cycle from CE stage. Questioning to elaborate those micro mistakes from past can serve the purpose of RO stage which results in development of some compact understanding towards the consequences in correlation of AC. Playing this final level of LABSCAPE precisely terms AE stage to end this cycle.

Cycle IV: Final cycle of ELT spiral where a game player enters to real laboratory to perform the Nitration reaction carrying the existing knowledge gained from LABSCAPE I to III levels for CE stage. Discussing game levels with other students in laboratory serves RO purpose with the output of AC to perform actual laboratory experiment (AE). This ELT spiral ends here with further learning cycles, where the base of the ELT would start from this level.

Conclusion

Mobile phone applications are increasingly in use for engineering education and chemical industry for different purposes. This paper was aimed to highlight the early stage progress of how we believe mobile game application could probably work on replacement of traditional safety training for senior staff or students who is working in same laboratory since long. Considering potential hazards of using mobile phone in laboratory, even if it has been suggested to play LABSCAPE before or after laboratory hours, it may be possible of continuity of LABSCAPE during laboratory hours. This would be conflicting towards the overall aim of this game with creating hazardous situation by playing LABSCAPE.

References

[1] J.Lu, Z.Meng, and J.B. Stav, "A new approach in improving operational efficiency of wireless response system," Computer and Information Technology (CIT), 2010 IEEE 10th International Conference on IEEE, (2010) 2676-2683

[2] C.Franzwa,Y. Tang, and A. Johnson, "Serious Game Design: Motivating Students through a Balance of Fun and Learning," Games and Virtual Worlds for Serious Applications (VS-GAMES), 5th International Conference on IEEE (2013), 1-7.

[3] D. Pano, C. Baulch, Smartphone use in process safety, Available web link with https://onlinelibrary.wiley.com/doi/10.1002/prs.11884, Accessed on February 13, 2019

[4] J. A. Frank and V. Kapila, Development of Mobile Interfaces to Interact with Automatic Control Experiments,. IEEE Control Systems, 2014

[5] Volodymyr Oliennikov, OSHA Safety Regulations, Checklists Audit Reports, 2016.

[6] W. Howard, OSHA Safety, 2013, Available web link at https://itunes.apple.com/us/app/osha-safety/id690989492?mt=8, access on January 15, 2019.

[7] W. T. Dotehlo, M. Marietto, and J. C. M. Ferreira, Kolb's experiential learning theory and Belhot's learning cycle guiding the use of computer simulation in engineering education: A pedagogical proposal to shift toward an experiential pedagogy, Computer Applications in Engineering Education 24 (2015), 79-81.

[8] M. Gunasekera, F. Khan and S. Ahmed, Learning Process Safety principles through practice, Process Saf Prog 37 (2018), 347- 354.

[9] N. D. Dholakiya, M. Ferjencik, D. Schofield and J. Kubik, Virtual learning for safety, why not a smartphone? , Available web link with https://onlinelibrary.wiley.com/doi/10.1002/prs.12005, Accessed on February 13, 2019.

[10] D.A. Kolb, Experiential learning: experience as the source of learning and development, Prentice-Hall, Englewood Cliffs, New Jersey, 1984.

[11] W. T. Dotehlo, M. Marietto, and J. C. M. Ferreira, Kolb's experiential learning theory and Belhot's learning cycle guiding the use of computer simulation in engineering

education: A pedagogical proposal to shift toward an experiential pedagogy, Computer Applications in Engineering Education 24 (2015), 79-81.

[12] D.A. Kolb, Experiential learning: experience as the source of learning and development, Prentice-Hall, Englewood Cliffs, New Jersey, 1984.

546961 Leveraging Advanced Data Analytics to Improve Process and Plant Safety

Abstract:

Digitalization is the next step in the industrial revolution to drive improvements in productivity within the Process Industries. This is made possible by the incredible speed at which new technological driver's the industry has available such as Increased computing power, improved communications, smarter sensors, virtualized architectures, cloud computing and process simulation. Historically, the responsibility to reduce process risk by reducing the likelihood of hazardous event was placed on the Safety Instrumented System (SIS). Today, plants are starting to look at areas to improve their overall plants operations to additionally reduce the initiating events that might have lead into these hazardous conditions.

What attendees will learn

In this session attendees will learn how advanced data analytics in their process plants can help reduce potentially hazardous scenarios, thus lowering their overall process risk. Reducing process risk through these methods will lessen the performance requirements on their SIS design which will help to lower costs and improved plant availability. The following areas to leverage data analytics for improving process and plant safety will be explored:

1. Optimize control loops to reduce potentially critical situations.
2. Reducing stress and work load of the operator by improving alarm and messaging systems.
3. Avoid unexpected plant downtimes by predicting failures before they happen to key process assets (valves, motors, drives, etc.)

551236 Data-Assisted Optimization

In many engineering fields, there is a continuously increasing interest in coupling equation-based first-principle modeling with information that comes in the form of input-output data for optimization. This approach enables the optimization of systems incorporating very detailed information regarding the material, flow, geometry, physical properties and chemistry of the systems. This need has given rise to development of methods that can optimize systems without equations or derivatives, but simply through the exchange of input-output data streams. The typical applications of data-driven optimization consider the system under study entirely as a “black-box”, however, many applications exist (i.e., process synthesis and design of modular manufacturing systems) which can be formulated as hybrid problems. Hybrid systems are comprised of both explicitly known equations (first principles) and data-dependent equations. This talk will present techniques for efficient optimization of such hybrid problems. Specifically, we propose and compare several methods for developing adaptive, flexible and tractable surrogate parametric functions to represent the information that comes in the form of input-output data. Finally, we present a data-assisted algorithm for optimization of these hybrid formulations, which integrates concepts from mathematical programming with machine learning techniques for regression and feature selection.

550736 Predictive Modeling for Product Quality in Polymerization Process

Achieving consistent and on-spec product quality is critical for chemical production processes. Final product samples are taken at the end of processes to measure key product qualities in the lab. Polymerization processes have substantially long and varying residence time between process adjustments, product sampling and lab analysis, depending on production rates. The long lag time means that there is a high potential for long periods of sub-prime material production before getting back on-spec. Sub-prime material inventory reduces a plant's effective production capacity and incurs added cost for storage, segregation and blending, which negatively impacts working capital performance and profitability. Hence, there is a need to develop predictive analytic models that provide real-time prediction of final product quality, enable process adjustment without delay, and minimize the production of off-spec products.

This paper presents a Machine Learning-based predictive model developed by Honeywell Connected Plant for a chemical plant located in Texas to provide real-time prediction of final product viscosity. The prediction model was developed using process data, laboratory data, advanced data analytics coupled with deep process understanding. To prevent overfitting, Machine Learning leverages cross-validation, splitting the data set into a training data set and a validation data set to ensure models are both highly accurate and generalize well to new situations. It was found that the selection of relevant contributing variables and proper data aggregation and preparation are critical for developing an accurate and meaningful data-driven predictive model. The effects of model input variables are consistent with polymerization reaction mechanism. The final selected model successfully predicts viscosity for multiple product grades.

550285 Soft Sensor Development for Industrial Milk Powder Process Systems

The potential of "big data analytics", characterised by volume, velocity, variety, veracity and value has, not unexpectedly, attracted the attention of the Dairy Processing industry. While classical big data techniques have been successfully applied to dairy herd management, genetics, with the aim to increase milk production from cows, the application of big data mining techniques to industrial process data is still relatively new.

One of the most important questions to ask is what you want to learn from taking a big data approach. Most big data applications focus on the 'what', trying to find trends or clusters in what is occurring without necessarily understanding the 'why'. However, adding the 'why' into the equation changes the usage of big data landscape, and this is what we are interested in, in the milk powder processing industry. For example, why is powder occasionally off-specification on particular properties.

Industrial processing plants typically collect huge volumes of data, much of which is dumped to a historian, never to see the light of day. From our experience and the literature, the problem is that the big data value is extremely low. The big data does not contain enough or very little key process/quality variables strongly linking to the final product quality.

To address the above question, we developed a soft sensor which can predict a milk powder quality attribute, dispersibility, based on its morphology, so that the big data value is enriched. The functionality and performance of instant whole milk powder (IWMP) can be related to its physical and chemical properties such as particle size distribution (PSD), morphology and fat distribution etc. The dispersibility, a measure of rehydration, has become one of the most important quality characteristics of IWMP. Many milk powder properties may affect the dispersibility, based on the literature and our own experience, PSD and morphology may be the two most important factors. A soft sensor based on a quantitative relationship between the PSD/morphology and dispersibility was built using an artificial neural network. The results may help dairy industry to improve their milk powder online monitoring.

551667 Persistent Homology to Visualize Batch Data Sets in Chemical Processes

Visualizing big data sets in a reduced dimensional space is a key step for root cause identification in batch chemical processes. Persistent homology, a topological method, provides a novel way to summarize big data sets and link differences with process conditions by extracting characteristics of the high dimensional data space, such as the number of clusters. Visualization based on persistent homology brings similar information and requires less preprocessing of the data, such as an alignment, when comparing with traditional PCA and PLS multi-way approaches. The performance of the persistent homology method is also compared with UMAP (Uniform Manifold Approximation and Projection for Dimension Reduction) and t-sne (t-Distributed Stochastic Neighbor Embedding), demonstrating that persistent homology requires less hyper parameters for tuning. An industrial case study for chlorine production was utilized to evaluate performance of the visualization methods, where the source of run length underperformance was determined by examining specific process conditions during 7 years of operation.

550484 Manufacturing Smart Ideas -- New Ideas on Moving R&D Innovations to Manufacturing Faster and at Higher Value

The purpose of this talk is to frame the moving of R&D-to-manufacturing as a type of "chemical process". That is, we first must "manufacture the innovative ideas and designs", before we manufacture the products themselves. Framing new product development (NPD) innovations in manufacturing language and measurements enables dramatic improvements in NPD or development throughput, value of products and services for customers, speed from idea to market, and improved sustainability. In this talk we analyze the innovation process at three levels: 1) The system level (e.g., operations like stage-gates, business cases, R&D, regulatory issues), 2) the flow level (e.g., customer needs, potential inputs, recycles and purges of ideas), and 3) the "knowlecular level" (e.g., details like the particular questions, hypotheses, measurements used). These three levels have important similarities to how we analyze chemical plants more generally. This talk will show a dollarization of the benefits, and provide specific recommendations for action.

550525 Machine Learning Techniques Applied to Classical Product Properties

In recent years, the increasing integration of the Internet of Things into production industry is at the genesis of a new digital industrial revolution known as Industry 4.0 [1]. The core component of Industry 4.0 is the concept of the digital twin. The main objectives of digital twin are to replace or reduce expensive, time-consuming physical experiments with rapid, inexpensive computer simulation [2].

Accurate and reliable predictive models for physicochemical properties of Hydrocracking process (HCK) products are extremely important. It can help petroleum refinery industries to save time and expansion on costly experiments. In our case, this is the main motivation to synthesize the available knowledge base of HCK process to build a digital twin capable of predicting the product properties of valuable petroleum fractions based on scientific principles. However, for the efficient execution of digital twins is it required to use the different steps of the Knowledge Discovery in Databases-process (KDD-process) [3]. That means automatic extraction of non-obvious, hidden knowledge from large volumes of data [4]. Ideally, the twin would enable:

1. **Data cleaning and preprocessing** to handle:
    - missing data items by deletion or imputation approach,
    - unexpected values (variables under consideration are expected to have values within a predefined range) by expert pre-processing.
2. **Outliers Detection** to identify and remove unwanted samples from data. In this work, we use the Local Outlier Factor (LOF) technique [5].
3. **Variables selection** to select optimum number of variables from a large pool of variables. In the present work we applied leaps [6] and Random Forests [7] algorithms to determine suitable descriptors.
4. **Machine learning** to build models that characterize the impact of different physicochemical properties of the product. For this, Linear Regression, Kriging ,Support Vector Regression , Random Forest and Gradient Boosting Machine are proposed
5. **Validation** of the models using either a dedicated database or cross-validation techniques.

The proposed twin was tested to predict the specific gravity of the vacuum gasoil (VGO), diesel and heavy naphtha cuts in Mild HCK (process that uses low to intermediary pressures and relatively low conversions). The results from this work will be presented. Results are promising but many unforeseen difficulties had to be addressed.

In the future, the proposed framework can also be used to predict other properties (cetane number, sulphur and nitrogen content, etc.) with several process (High Pressure Hydrocracking, FCC …). Furthermore, this type of methodology is also extended to predict industrial deactivation.

References

[1] F. Shrouf, J. Ordieres, G. Miragliotta. Smart factories in Industry 4.0: A review of the concept and of energy management approached in production based on the Internet of Things paradigm. *2014 IEEE International Conference on Industrial Engineering and Engineering Management*, 2014, 697-701.

[2] Grieves M., Vickers J. Digital Twin: Mitigating Unpredictable, Undesirable Emergent Behavior in Complex Systems, in *Transdisciplinary Perspectives on Complex Systems: New Findings and Approaches*. Éd. F.-J. Kahlen, S. Flumerfelt, A. Alves. Springer International Publishing, Cham, 2017, 85-113.

[3] Schuh G., Rudolf S., Riesener M., others. Design for Industrie 4.0. *2016 14th International Design Conference*, 1387-1396.

[4] Fayyad U., Piatetsky-Shapiro G., Smyth P. The KDD Process for Extracting Useful Knowledge from Volumes of Data, *Commun. ACM*, 1996, **39,** 11, 27-34. DOI: 10.1145/240455.240464.

[5] Ding H., Ding K., Zhang J., Wang Y., Gao L., Li Y., Chen F., Shao Z., Lai W. Local outlier factor-based fault detection and evaluation of photovoltaic system, *Solar Energy*, 2018, **164**, 139-148. DOI: 10.1016/j.solener.2018.01.049.

[6] Feorge M. Furnival, Robert W. Wilson. Regressions by Leaps and Bounds, 16, **1974.**

[7] Genuer R., Poggi J.-M., Tuleau-Malot C., Villa-Vialaneix N. Random Forests for Big Data, *Big Data Research*, 2017, **9**, 28-46. DOI: 10.1016/j.bdr.2017.07.003.

551242 Alarm Management Using Data-Driven Methods

Alarm systems are a crucial part of modern control systems. They serve as a layer of protection and provide a prior indication to the operator about an abnormal process event. Alarms are designed and configured in the control systems and logged into historian for future references and reviews. With the progress in data mining and analysis methods, the alarm and event logs can be mined to generate diagnostic information about the healthiness and configuration of the alarm system. To demonstrate this, we propose a method to mine information from alarm & event log. This method is developed on an open-source programming platform. The input to the tool is an alarm and event log file from an asset management system or control system. This tool aids in (1) identifying and extracting alarm flood sequences, (2) providing graphs for visual analysis, (3) mining frequent alarm patterns and (4) bench marking the system against the Key Performance Indicators (KPI). The outputs can be used for offline analysis, rationalization of the alarm system, definition of suppression logics, and reduction in the alarm floods. The effectiveness of the proposed method and tool is established by application to a real industrial data set.

546373 Finding Unknown Anomalies before They Risk Process Safety

Four out of every five oil historical refinery accidents resulted from unexpected, unknown process and mechanical anomalies rather than known, expected failure modes. These unknown anomalies caused unexpected expenses of more than $100 million and ruined excellent company safety records. The oil refinery industry works diligently to address findings from all accidents to avoid repeat incidents. However, the next unknown anomaly which may cause a large property damage loss lurks obscured in a fog of maintenance and operations variance. Rather than wait fearfully for the next unknown anomaly to surface, AI Factory detects all known and unknowns as early as possible. This paper will explain how the latest Artificial Intelligence and Deep Learning techniques from BHGE, a GE company, help eliminate operational surprises. Because refineries generate a wealth of rich data, they need a fundamentally different approach to use information to detect anomalies which lead to 9 digit accidents. Using AI Factory, refiners will find unknown anomalies early and avoid catastrophic surprises. In addition, they can now rapidly refine data driven models and visualize unique, industrial scale operations.

546723 Digitalization in Process Safety: Fire Risk Prediction Analytics

As PETRONAS move towards digitalization, we envision to converge HSSE and operational excellence and realized the value of our extensive process data. One of our focus is to identify emerging process safety risk utilizing process safety data. With that, we embark on a pilot analytics project with the aim to enable timely interventions to prevent process safety incidents.

The analytic model is build based on Data-driven model and SME-input model. Data-driven model learns the pattern of past incidents while the SME input model detects abnormality based on available process safety information. Weightage is assigned to process indicators and alarm excursions to obtain a risk score and provide an early warning for intervention.

Furthermore, skeptical analysis was conducted against process safety leading indicators determine if there are any early warning indicators prior to an incident.

The pilot model was run on 5-years data in one of our facility with a selected recall accuracy and prediction accuracy. This set the basis of minimum viable product. Real time monitoring is currently in place for the pilot model. The results will serve as a basis for future risk analytic projects.

543515 Big Data, Iiot, & Digitalization: That's the Answer, What's the Question

This session will touch the key initiatives in industrial IT: Digitalization, BIG DATA, APM, IIoT, Cloud, Edge, etc., and explores the reasons to pursue them; and how not to. It goes on to explain and explore the single most important: maximizing uptime (availability to plan). Probing past and contemporary maintenance practices that may or may not have assisted uptime will identify a new, assertive direction using analytical techniques. There is a lot of talk these days about analytics and machine learning. It's everywhere. Is it all the same? You can get it in kits from Microsoft with Azure, Google with Tensor Flow, MATLAB, and others. But not all analytics and machine learning are inherently designed for asset protection. It's not the same as credit card fraud detection, or facial recognition, and bears significant consequences when it's wrong. What does it really take to create a world that doesn't break down? Vendors offer the same buzzwords all products sound the same. This session identifies the key capabilities to explain the most important. That means not just reducing maintenance costs but maximizing uptime through extreme early warnings that allow planning to fix small problems before they become catastrophic. Not just late-stage damage alarms but identifying the upstream root causes of damage and advising process condition adjustment to eliminate the causes of the most failures. Your operators and maintenance staff can then be confident in the failure prediction and act together to avoid unplanned production downtime.

550047 Data Analytics Gives Manufacturing Plants Insights for Continuous Improvement

Industry 4.0 and the Industrial Internet of Things (IIOT) promise to make more data and better analytic applications available for industry. Driven by high powered analytics, including machine learning and artificial intelligence engines, both promise operational benefits including lower costs through improved Key Performance Indicators (KPI), such as First Pass Yield, Non-Standard Downtime or Overall Equipment Effectiveness. However, despite the investments in data capture and analytics, the industry faces a daunting challenge when attempting to drive continuous improvement, given the ***amount, types and lack of context*** in most manufacturing data. The result on the shop floor is often characterized by data overload and inaction. We present a proven strategy for continuous improvement in manufacturing that can provide (1) insights into process variability, (2) correlations with product quality measures, (3) real-time statistical process control (SPC), (4) hierarchically structured data for regulatory compliance, and (5) powerful machine learning algorithms to handle massive and disparate data common in process industries. Incorporation of this strategy into Continuous Improvement systems (Lean Six Sigma) has led to bottom line improvements in hundreds of work centers in a global infrastructure. An example from a Chemical Reactor is presented to illustrate the methodology and application in both an on premise and cloud based environment. Techniques are utilized employing both univariate and machine learning algorithms to illustrate how to explore and monitor complex manufacturing data sets.

550387 Ready for Digitalization - Don't Miss the Train

Whether a brownfield or greenfield facility, a company manufacturing chemicals or food supplements or a 10 person startup versus a Fortune 500 company, businesses of all sizes and sectors can achieve a simple entry into the world of digital transformation. Depending on the degree of digitalization, users are able to implement either the full scope of the Digital Enterprise or simply pick individual digital applications that can get them started on their digital journey.

One of the focal points of digitalization is the ability to increase the availability of plant assets and to give on-site personnel quick and meaningful decision-making tools for their work. The objective is to make processing plants more reliable through proven products and innovative services and to be able to more dependably predict the condition of assets. With the dawn of new and innovative analytical methods powered by secure, open, cloud-based IoT operating systems, significant amounts of data is collected, put into a meaningful context and intelligently linked in data models. Smart data applications provide support for all levels of the organization. They enable potential asset or component defects to be predicted, as well as helping to minimize plant downtime.

The overall goal of the presentation will be to provide a solid understanding of the power of digitalization and gain insights into how to get onboard the digitalization train. Some of the concepts to be discussed in more detail include:

- The digital twin, 3D and simulation and how it impacts automation and digitalization across engineering, operations and maintenance
- The increasing importance of cybersecurity
- Innovative ways of gathering data at the field level to be used for gaining new insights into the process
- How to bring plant and tag data into the field and to make optimal use of remote service capabilities to enable faster and more efficient maintenance
- Provide specific examples of how to use existing asset data to provide insights that provide value in terms of improved predictive maintenance, less downtime and faster maintenance

549503 Best Big Data Analytics Platform for Manufacturers

Artificial Intelligence, Machine Learning, and the Industrial Internet of Things are still the hot topics in the big data analytics world. Manufacturers have been generating data off the line for decades, so how can manufacturers get the most of these new advancements in analytics? The combination and convergence of new hardware infrastructures have allowed for the advancement of analytics like artificial intelligence to move from research out into the business world. Analytic insights need to be deployed into production and actions need to be taken sometimes in real-time to truly realize the benefits of all these advancements. Come see how the power of the SAS platform with it's support of the entire analytics lifecycle: Data, Discovery, and Deployment provides the best big data analytics platform for manufacturers to leverage all sorts of analytics both in batch and in real-time.

548869 An Holistic Approach for Batch Analysis and Monitoring

ABB Ability™ BatchInsight identifies and monitors anomalies in the batch process. It helps process specialists and plant operators detect and resolve abnormalities by conducting batch-level analysis.

Using time-level analysis, it recognizes the time of occurrence of the deviation and isolates the process variables that contributed to the deviation, enabling effective and timely corrective action.

The big data algorithm empowered batch monitor supervises continuously the running batch process. It enriches a normal PV monitor with additional features like, similarity-based process forecasts, adaptive "golden batch" process corridor, batch-phase change recognition & visualization, and operator decision support after early warning.

550737 Predictive Maintenance with Machine Learning

In the process industries, many companies are trying to reduce unexpected failures in their equipment by developing predictive maintenance algorithms. Some of the difficulties associated with developing these algorithms include the lack of failure data from the machine, as well as being unable to incorporate their engineering and modeling expertise into the final solution.

In this session, MathWorks will show you how to use physical models of your system to generate failure data by simulating fault conditions, and then using this data to train a predictive model in MATLAB. We also show you how to integrate these algorithms with your embedded devices so that your algorithms operate directly on real data coming from your machine. Specifically, we will concentrate on a triplex reciprocating pump – an example that was requested by several industry practitioners.

549955 Heat Exchanger Fouling and Cleaning Analysis for Complex Network

Heat exchanger fouling has a direct impact on plant profitability. To combat fouling, heat exchangers must be periodically removed from service and cleaned. Frequent cleaning substantially increases cleaning expenditures. However, cleaning too infrequently has the potential to increase energy consumption due to limited heat transfer capabilities of the fouled heat exchangers. In both cases, operating cost goes up.

Many refineries conduct cleaning on a pre-determined schedule. Some only perform cleaning when operating problems, such as hydraulic restrictions or unacceptably high rundown temperatures, become apparent. Others do periodic monitoring of individual heat exchanger performance to identify the most badly fouled heat exchangers. However, a heat exchanger is not necessarily worth cleaning simply because it is heavily fouled. In case of a crude unit, heat exchangers at the front-end of the preheat train have very limited impact on the temperature of the crude oil by the time it reaches the atmospheric heater, as a loss of heat recovery in an upstream heat exchanger is usually offset by increases in heat recovery in downstream heat exchangers. Hence, in addition to individual heat exchanger monitoring, it is also necessary to analyze the knock-on effects throughout the heat exchanger network.

This paper presents heat exchanger fouling and cleaning analysis of a crude preheat train with the objective of mitigating fouling effects through improved cleaning schedules. In this analysis, rigorous heat exchanger network simulation of a crude preheat train at various fouling conditions is performed in UniSim to account for heat exchanger interactions and identify critical heat exchangers in terms of fouling effects that need to be monitored closely. In addition, UniSim simulation results are coupled with economic factors, other asset conditions and operational scenarios to optimize cleaning schedules resulting in overall economic savings (OPEX reduction). The findings from this analysis also provide valuable insights for maintenance scheduling, turnaround planning, and process improvement.

550592 Performance Evaluation of Alternative Technologies for Fouling Mitigation in Refinery Pre-Heat Trains

While heat exchanger fouling management tools can be effectively used in refineries to minimise the impact of deposition from crude oils, it is desirable to mitigate fouling by design. This has been a challenge for many years in part because of the lack of understanding of the fundamental mechanisms leading to deposition and in part for the intrinsic limitations of the methodologies traditionally used for heat exchanger design. Moreover, opportunities to use non-conventional mitigation technologies (e.g. helical baffles, welded plates, tube inserts) are often missed out, given the challenges in evaluating their performance with respect to traditional shell-and-tube heat exchangers.

In this paper, a methodology to evaluate the performance of different technologies for fouling mitigation with plant data is demonstrated. The hot end of a refinery preheat train is analysed to identify for retrofit opportunities to mitigate fouling. Several retrofit options are proposed for the heat exchanger with greater potential for improvement. By using fouling predictive models, previously fitted and validated using plant measurements, it is possible to re-run history and assess how alternative designs would have performed. The evaluation includes the evolution of fouling, the time-variation of the thermal and hydraulic performance of the heat exchanger retrofitted, the network interactions, and, ultimately, the economic savings that would have been achieved if the alternative technologies had been placed instead of the original one. The methodology is shown to provide a quantitative assessment of the benefits and trade-offs involved in the retrofit, de-risking the selection of non-conventional technology.

Proceedings 544039

Crude Unit Fouling: Consequences and Costs

Andrew W. Sloley

Advisian
WorleyParsons Group

Advisian (WorleyParsons Group)
Energy Center II
575 N Dairy Ashford St
Houston, TX 77079

Prepared for Presentation at the
American Institute of Chemical Engineers
Spring Meeting
New Orleans, LA, March 31 – April 4, 2019

UNPUBLISHED

1. Introduction

Most crude unit preheat trains use shell-and-tube exchangers for their heat recovery. Heat exchanger fouling adds considerable cost to both construction and operation of crude units. Higher fouling shifts optimum heat recovery investment to higher capital and lower energy recovery. The industry has taken three approaches to account for fouling in heat exchanger design:

- Fouling factors
- Over-surface
- Cleanliness factors

Fouling factors add a resistance term to the heat-transfer calculations. This factor can be applied to either or both sides of the service. Using a fouling factor is the most common industrial practice. It is used extensively to design exchangers, evaluate performance, and determine exchanger cleaning schedules.

The second approach uses an over-surface specification. The exchanger is designed as if it were clean, then extra area is added. This approach has the benefit of being explicit in the cost of the overdesign but is much more difficult to use in optimizing cleaning schedules of existing equipment.

The cleanliness factor is a de-rating factor applied to the overall heat-transfer coefficient. It functions similarly to an over-surface margin specified when designing new equipment. The cleanliness factor approach is most often used in the design of surface condensers in the power industry. In that service, overall heat-transfer coefficients are high and nearly all the fouling is on the cooling water (tube) side of the exchangers.

Spirited discussion continues on which method is best for the design and evaluation of crude unit exchangers. The exchangers are well known for severe and persistent fouling, a problem that has defeated attempts at solution for decades. Using too low a fouling level leads to units that cannot meet energy recovery, capacity, or quality targets. Using too high a fouling level leads to excessive investment in heat exchangers, and may have little-to-no performance improvement if done incorrectly.

This paper examines both the fouling-factor approach and the over-surface approach and compares them. Data are presented from four operating crude units. All four of the units exclusively use shell-and-tube exchangers to recover energy for heating the crude oil feed. Previous work [1] based on a literature review identified reasonable fouling factors for commercial units, even though fouling may vary dramatically from unit to unit and from exchanger to exchanger within the same unit.

This study answers the following questions:

A. What is the relationship (if any) between fouling factors and an equivalent over-surface percent?
B. What is the heat recovery cost of actual versus TEMA[a] fouling factors?
C. What is a realistic over-surface percent or fouling factor for a crude unit?

2. Factors Affecting Fouling

Regardless of the method used to account for, or alleviate, crude unit fouling, fouling in crude units has been a persistent problem for decades. One of the author's first experiences in his first year of working was to review a colleague's progress in predicting/preventing fouling of crude unit exchangers. There hadn't been any progress for years. Every idea and theory had failed.

The fundamental issue is that crude fouling is complex. It arises from a combination of the following factors:

- Crudes and how their blending properties interact, leading to colloidal and composition interactions
- Crude history including blending order and exposure to oxygen and other chemicals
- Desalting and dehydration performance in the tank farm and in the crude unit
- Composition and structure of the residual mineral content
- Fouling deposit aging, crude deposit reactions, and other factors
- Shear stress (velocity) in the exchanger
- Exchanger configuration including fluid allocation, type, and dead spots in the flow patterns
- Differences between surface, film, and bulk temperatures, and differences in properties arising from them
- Differences in the multiple fouling mechanisms present including asphaltene precipitation, carbonate formation, wax deposition, chlorides and other salt deposition, contained solids deposition, and fouling-corrosion interaction

Along with this comes the difficulty in gathering high-quality data, and uncertainty in correlation accuracy. In operating plants it is impossible to separate the effects of fouling on one side of the exchanger from the effects of fouling on the other side. Also, operating plants may suffer from incomplete exchanger cleaning and exchanger damage. Both of these confuse data analysis.

In the industry as a whole, progress has been made. The progress has been by small, cumulative gains in exchanger design, chemical additives, desalting improvement, and crude blending knowledge. However, large step changes that would prevent energy and capacity losses due to fouling have eluded the industry.

[a] TEMA, Tubular Exchanger Manufacturers Association

3. Fouling Factors

3.1. What are Fouling Factors?

Fouling factors add a thermal resistance to the overall heat-transfer calculation. Equation 1 shows the overall heat-transfer coefficient (U) as a function of the tubes' outside heat-transfer coefficient (h_o), resistance to heat transfer of the outside tube surface (r_o), resistance of the wall (r_w), resistance of the inside tube surface (r_i), inside heat-transfer coefficient (h_i), and surface areas of the outside of the tube (A_o) and inside of the tube (A_i). A fouling factor is used to show the thermal resistance to fouling, giving a value to either r_o or r_i.

Equation 1

$$U = \frac{1}{\left[\frac{1}{h_o} + r_o + r_w + r_i\frac{A_o}{A_i} + \frac{1}{h_i}\frac{A_0}{A_i}\right]}$$

In design, fouling should represent fouling alone. The fouling factor should not be used to substitute for uncertainties in correlations and physical property predictions.

The best source of information on fouling factors should be experience with similar exchangers on similar feeds. However, at the design level this information is rarely available.

TEMA publishes 'Recommended Good Practice' values in their standards [2]. With few updates, these values are mostly the same as they were when originally published in 1941. No documentation exists supporting the original definition of the fouling factors. Based on a summary of discussion with early members of TEMA, the conclusion is that the original values were a *"consensus drawn from many years of practical experience"* [3].

TEMA factors include values for both salted crude (containing produced water with soluble salts), and dry crude. Fouling occurs on both sides of exchangers. TEMA includes factors for both the crude and the heat recovery streams (crude unit products and heat removal streams). The analysis here looks at a total fouling factor of tubeside fouling plus shellside fouling. Based on TEMA, most crude exchangers should be designed with a total fouling factor between 0.004 and 0.007 ft^2-°F-hr/BTU (0.0007-0.0012 m^2-°C/W). For heat integration with residue streams, TEMA recommends total fouling values of up to 0.014 ft^2-°F-hr (0.0025 m^2-°C/W).

3.2. Advantages of Using Fouling Factors

Fouling factors have long been used in the industry. Whether they are accurate or not, the values published by TEMA find common acceptance for design purposes. They also have a fundamental place in the overall heat transfer in Equation 1. How the fouling resistance shows up is clear and transparent. The use of fouling factors allows heat-transfer engineers to optimize exchanger design for specific fouling services. Table 1 shows common parameters that may be used to optimize the exchanger.

Table 1 Exchanger Optimization Parameters

Shell-Side Optimization Features	Tube-Side Optimization Features
Which fluid in shell (fluid allocation)	Which fluid in tube (fluid allocation)
Configuration	Tube passes
Diameter, length	Tube diameter
Baffle type, cut, spacing	Tube length
Tube spacing	Tube count
Tube pattern	Pressure drop
Pressure drop	

TEMA fouling factors are accepted by many as de-facto standards. In recent years, other sources [4] have given examples of fouling factors in specific applications, but little published work gives overall recommendations.

For monitoring, most approaches (and commercial software) calculate a fouling factor based on measured flow rates and temperatures, calculated physical properties, exchanger configuration information, and a selected method for rigorous evaluation of the heat exchanger. The difference between expected clean performance and actual measured performance is balanced by finding the fouling factor that accounts for the difference.

In monitoring, the fouling factor combines fouling, data errors, calculation inaccuracy, problems due to exchanger damage, and any other contributing factors. Because of this, nominally clean exchangers may give results that imply extensive fouling. In other cases, exchangers may even have negative fouling factors.

3.3. Disadvantages of Using Fouling Factors

Designing with fouling factors has three major disadvantages.

First, the fouling factor can obscure how much extra surface area is added to the exchanger to maintain performance when fouled. Often, the cost implications of fouling factors are not well understood. This affects both the design of specific exchangers as well as the overall design and optimization of entire preheat trains.

Second, when fouling factors are high, exchanger areas become large. Large exchangers tend to have relatively low fluid velocities at the optimum design point. Extensive industry experience shows that low fluid velocity (and thus low shear stress) increases fouling rates [5]. The exact relationship can be disputed, but the trends are very clear.

Third, fouling factors are fixed. Fouling is dynamic. In some cases, reliable data have shown that under some conditions for some exchangers, there are periods when fouling decreases. Also, specific fouling values are based on assumptions about cleaning schedules and practices—which are unknown when designing the unit.

3.4. Unimportant Considerations

A fourth criticism is sometimes leveled at fouling factors, namely that their values are too low. That may be true, but this criticism targets data adequacy, not a fundamental fault with the method or its application.

On the other hand, an advantage often quoted for fouling factors is that they explicitly allow designs to account for differences in fouling tendencies of different streams. This has little real value for design, for two reasons. In most crude exchangers, the dominant fouling is on the crude side. Also, when fouling factors are derived from operating plant data, the overall value is split arbitrarily and allocated to one side of the exchanger or the other. Stream fouling factors are thus little more than educated guesses.

3.5. Reasonable Fouling Factors

Defining a reasonable fouling factor requires looking at two things: first, how much fouling happens, and second, is the analysis looking at a single exchanger or an entire preheat train?

As previously noted, using TEMA values will generally result in crude exchangers with total fouling factors from 0.004 to 0.007 ft^2-°F-hr/BTU (0.0007-0.0012 m^2-°C/W).

Based on a late 1980's survey [3], about 70% of new crude preheat exchangers had a specified total fouling factor of 0.004 to 0.006 ft^2-°F-hr/BTU (0.0007-0.0011 m^2-°C/W). However, some exchangers were designed with fouling factors specified as high as 0.030 ft^2-°F-hr/BTU (0.0053 m^2-°C/W), five to eight times the normally accepted value.

Recent review of published data by the author examined 139 exchangers from multiple units [1]. These data are mainly from 1990-2015 and are biased toward problem cases. Fouling followed a bimodal distribution, with one set of low-fouling (normal) exchangers and another set of high-fouling (problem) exchangers. The important parts of this analysis show:

- For the problem exchangers, the mean fouling level when nominally clean at startup exceeded the TEMA values.
- The dominant pattern of fouling in crude exchangers over time is a linear increase, though the fouling can be subject to changes in slope, or display sudden steps.
- For the low-fouling exchangers, the fouling factor seen at end-of-run (or just before cleaning) was 2.4-2.8 times the TEMA values. Forty-nine exchangers were in this group.
- For the high-fouling exchangers, the fouling factor seen at end-of-run (or just before cleaning) was 6.5-8.1 times the TEMA values.
- No correlation was seen for overall fouling factor versus the nature of the stream on the non-crude side of the exchanger.

What constitutes a reasonable fouling factor depends upon the consequences of using the wrong factor. When looking at an entire preheat train, not every exchanger will be at maximum fouling at the same time. Some exchangers will be fouled, some will have just been cleaned, some will be fouling faster than average, some will be fouling slower than average. Additionally, heat not recovered in one exchanger may be recovered by a downstream exchanger.

When looking at a critical service with a single exchanger, the fouling factor should meet the expected end-of-run conditions for that exchanger. For example, in Figure 1, exchanger service EX15A-C is the only exchanger in the kerosene pumparound circuit. If this exchanger transfers less heat than expected, all the lost heat goes to the tower condensers. The increased overhead duty could cause the tower to pressure up (condensers are now too small) or the tower to flood (too high a vapor load in the tower).

The final exchangers in the preheat train are most critical to the overall train performance. There are fewer, or no, exchangers downstream to compensate if they have poor heat recovery. In cases where the downstream heater may limit operation, these should be considered critical service exchangers.

Based on the previous work the author recommends the following fouling factors for design:

- Normal services – 0.0075 ft^2-°F-hr/BTU (0.0013 m^2-°C/W). This is 50% of the end-of-run or just-before-cleaning value for the heat exchanger.
- Critical services – 0.015-0.020 ft^2-°F-hr/BTU (0.0026-0.0035 m^2-°C/W)
- Final services – 0.015--0.020 ft^2-°F-hr/BTU (0.0026-0.0035 m^2-°C/W)

These factors are based on observed plant performance at typical 180-day to 250-day cleaning intervals.

4. Over-surface Factors

Over-surface specifies that a set fraction of extra area should be added to a heat exchanger. For design, the advantage of this approach is that it makes explicit the area (and cost) added to compensate for fouling. When using an over-surface approach, the design engineer should specify whether to add area by increasing diameter or by increasing length.

The preferred approach would be to add exchanger length. Pressure drop limits should be relaxed when factoring in extra surface. The hydraulics in the overall heat train should have enough allowance to account for the extra pressure drop.

In real units (rather than paper studies) most plants only allow specific standardized exchanger lengths for heat exchangers. This makes unit layout simpler and reduces costs by standardizing replacement tubes. If the specifications are set so that one standard length replaces another standard length (i.e. 16-ft long exchanger plus 25% over-surface gives a 20-ft long exchanger) this approach works well.

The alternative is to design exchangers to maintain a specified minimum fluid velocity when the over-surface is added. This allows the designer some flexibility in fitting an exchanger within standard dimensions. Again, pressure drop specifications should be flexible enough to account for the extra pressure drop required to meet the velocity specification.

The greatest problem with specifying an overdesign is that few engineers have any feel for how much is needed. When discussed, over-design values of 25-30% are often mentioned. These values are usually arbitrary and not based on sound economic analysis or verified data.

5. Operating Unit Comparisons

Data from test runs of four operating units have been analyzed to examine the consequences of fouling. None of these data were included in the individual exchangers reviewed in the previous work [1]. The units were chosen for the following major reasons:

Unit 1: A unit configured with heat integration with reboiled sidestrippers. This is an unusual configuration but has benefits for high yields of adjacent products with large differential values.

Unit 2: The most heat-integrated crude unit the author has worked on. This is a high-temperature topping unit (atmospheric tower only).

Unit 3: A unit that has crude on the shell side of all its crude preheat exchangers. Only approximately 30% of crude preheat exchangers are configured this way, and not much data has been published on fouling for these cases.

Unit 4: A unit with two nearly identical parallel trains.

Table 2 summarizes the performance and configuration of these units.

Table 2 shows three outlet temperatures for each of the preheat trains. The first temperature (Actual) is the measured outlet temperature during the unit test run. The second temperature (TEMA) is the predicted outlet temperature if the exchangers were fouled at TEMA values. The third temperature (Clean) is the predicted outlet temperature if all the exchangers were clean.

Unit 3 had just reached one year of operation at the time of the test run. The other units had been operating for two years or longer. The exchangers were all in what would be considered normal cleaning cycles. Unit 3 is the only one that exceeded the heat recovery expected with TEMA levels of fouling. That may be due to the shorter run before data were gathered.

The four operating units included 105 heat exchangers in crude preheat service. Over 95% of the exchangers were operating with crude velocities >8 ft/sec (2.4 m/sec). This high velocity would generally be thought to contribute to lower fouling. In spite of this, three units showed performance that indicated fouling was higher than TEMA-recommended levels.

Actual fouling on exchangers in operating units depends upon how often the exchangers are cleaned. The greatest cost of fouling comes from lost capacity and possible lost yields. Energy costs are rarely a significant factor compared to the capacity and yield losses. This is true even when energy costs are high.

Table 2 Configuration and Performance of Operating Units Examined

	Unit 1	Unit 2	Unit 3	Unit 4
Crude In, °F (°C)	71 (22)	70 (21)	94 (34)	105 (40)
Crude Out, ° F (°C)				
Actual	424 (218)	572 (300)	442 (228)	477 (247)
TEMA	435 (224)	609 (320)	431 (222)	540 (282)
Clean	463 (239)	648 (342)	455 (235)	570 (299)
Exchanger FFs	ft^2-°F-hr/BTU	ft^2-°F-hr/BTU	ft^2-°F-hr/BTU	ft^2-°F-hr/BTU
Maximum	0.0132	0.0609	0.0100	0.123
Minimum	0.0014	0.0030	0.0000	0.000
Exchanger FFs	m^2-°C/W	m^2-°C/W	m^2-°C/W	m^2-°C/W
Maximum	0.0023	0.0107	0.0018	0.0217
Minium	0.0002	0.0005	0.0000	0.0000
Configuration	Atm-vacuum	Topping with preflash tower	Atm-vac with preflash drum	Atm-vacuum
Why included	Reboiled sidestrippers	Most integrated	Crude on shellside	Parallel trains
Services, total	8	18	9	17
Services, crude on shell	2	1	9	2
Services, crude on tube	6	17	0	15
Exchangers	10	41	12	42
Notes		1, 2	3	2

1: Clean performance limited by premature crude vaporization.
2: Clean performance limited by hot-outlet–cold-outlet pinch (HOCO).
3: No desalter in unit.

The units examined are all for North American locations with low energy costs. For these units, over 90% of the cost of fouling comes from lost capacity and yield. Dramatically increasing energy costs would only slightly increase the total cost of fouling. Thus conclusions from this analysis are applicable to high-energy-cost locations with only minor caveats.

Of course, cleaning frequency varies with how rapidly an exchanger fouls. One interesting observation was that there was no correlation between in-service fouling factors when exchangers are cleaned versus operating temperature, heat recovery fluid, or whether the crude was desalted or not. This is because exchangers that foul more rapidly get cleaned more often. For the most part, fouling is not allowed to proceed past some maximum value.

For the four plants shown, the tolerated maximum overall fouling factor was about the same even though the value of lost capacity and the cost of energy in each plant were vastly different. The value is also very close to the value shown by the fouling factors in the previous survey [1]. This surprising conclusion is worth a short explanation.

The largest cost of fouling is lost production, even when energy prices are high. A fouled exchanger still causes production loss when removed from service. Cleaning economics are driven by time out of service (production loss) versus improved production after cleaning. By comparison, cleaning costs and energy efficiency—while not ignored—tend to be less important. Capacity value (money/capacity) is assumed to be equal for the production loss and gain.

Since most heat trains were designed using a limited range of fouling values taken from TEMA recommendations, they encompass a limited range of likely duty shifts as individual exchangers foul or un-foul (get cleaned). The duty shifts strongly affect the economics of exchanger cleaning.

In spite of the differences between plants, the combination of design parameters, operating conditions, and economic factors all come together to effect an almost identical maximum fouling factor across preheat trains and in most units. To follow the logic:

- all (or very nearly all) exchanger designs used TEMA-influenced fouling factors;
- within a single preheat train, all (or very nearly all) exchanger designs use the same target approach temperature between the crude and preheat stream;
- all (or very nearly all) plants are driven by capacity losses or gains;
- incremental losses and gains have equal value per unit production; and
- all (or very nearly all) fouling trends are roughly linear over reasonable time periods. [1]

Cleaning occurs when gains > losses. The exact value of energy does not matter very much because it's small relative to other costs. The exact value of unit capacity does not matter very much because it's equal on each side of the capacity tradeoff. While it is true that changes in external forces (energy economics) can be expected to change the fouling value at cleaning, it would take enormous shifts to produce a significant change. Life cycle analysis of energy is outside the current scope but the basic observation is that it takes a lot of energy to make the steel for the exchanger and piping. As the value of energy goes up, the value of the recovered energy goes up, but so does the energy cost of building and maintaining the preheat train.

Many factors are unknown during design, but operating results and the above analysis show that the unknown factors balance out when looking at entire preheat trains. Based on extensive data, the average fouling value at cleaning is 0.015 ft^2-°F-hr/BTU (0.0026 m^2-°C/W). The author does not see any likely energy cost change in the near future that would significantly change this analysis, and therefore recommends using one-half of this value for the average as a guideline for design (0.0075 ft^2-°F-hr/BTU or 0.013 m^2-°C/W).

(Note that this value is not meant as a criterion for deciding to clean specific exchangers. In operating units, cleaning decisions should be made on a case-by-case basis for every exchanger.)

6. Fouling Factors Observed

Figure 2 shows the distribution of fouling factors for Units 1-4. Figure 3 shows the histogram of services versus fouling factor ranges. The observed fouling factors ranged from zero to 0.123 ft^2-°F-hr/BTU (0.0217 m^2-°C/W). The median fouling factor was 0.0074 ft^2-°F-hr/BTU (0.0013 m^2-°C/W). The mean fouling factor was 0.0155 ft^2-°F-hr/BTU (0.0027 m^2-°C/W). These median/mean values are very close to the previously identified values of 0.0075 ft^2-°F-hr/BTU (0.0013 m^2-°C/W) for regular services and 0.015-0.020 ft^2-°F-hr/BTU (0.0026-0.0035 m^2-°C/W) for critical services.

These high fouling factors for the exchangers in the survey are observed in spite of much higher crude velocities in the exchanger (>8 ft/sec, >2.43 m/sec) than are called for in standard design practices (2-4 ft/sec, 0.61-1.21 m/sec). Although higher velocity does reduce fouling in most crude oil services, the data here show average fouling factors at high velocities equal to the upper range of TEMA fouling factors for lower velocities.

No trend was seen for reduced fouling on the shell side versus the tube side. The only fouling-motivated reason for putting the crude on a specific side of a shell-and-tube heat exchanger is the relative ease of cleaning one side versus the other.

Fouling factors included in the study showed values of up to 0.123 ft^2-°F-hr/BTU (0.0217 m^2-°C/W), roughly 18 times the TEMA value for this type of exchanger and service. How severe can fouling be in operating units? High quality plant data from a unit not included in this or the previous study resulted in a calculated fouling factor of 0.145 ft^2-°F-hr/BTU (0.0255 m^2-°C/W)—the highest fouling factor ever personally verified by the author. This is 29 times the TEMA value for the given service in spite of nearly monthly cleaning of the exchanger. In this case, the exchanger is a filter protecting other exchangers in the preheat train. Cleaning it has practically no benefit for heat transfer.

7. Over-surface versus Fouling Factors

All the exchangers in Units 1-4 were designed using fouling factors. For each service, an estimate of the percentage over-surface was generated for each of three situations:

1. TEMA fouling
2. Recommended fouling value (.0075 ft^2-°F-hr/BTU, 0.0013 m^2-°C/W)
3. Actual fouling

The over-surface number is the percent extra area required that would let the exchanger reach the design operating duty if the fouling level matched the fouling factor for that situation. In addition, a duty-averaged percent over-surface for the entire preheat train was estimated for the same situations. An extremely fouled exchanger does little duty, so the duty-average tends to damp the contribution of extremely fouled exchangers.

Attempting to create a comparison between fouling factors and over-design depends upon choices in optimizing between exchanger layout, pressure drop, fluid velocity in the exchanger, and other factors. Generally, it is thought that there is no strict correlation for fouling factor versus over-surface specification that would give identical heat exchangers. The approach used here was to lengthen or shorten the exchanger, allow the pressure drop to vary, and keep the velocities the same. Table 3 summarizes these results.

Table 3 Duty Averaged Over-surface Requirements

	Unit 1	Unit 2	Unit 3	Unit 4
	Over-surface percent			
TEMA Fouling Factor	58	60	38	70
0.0075 ft^2-°F-hr/BTU (0.0013 m^2-°C/W)	111	71	41	82
Actual Fouling Factor	87	122	18	187

Figure 4 compares over-surface for specific exchangers versus the fouling factor; it shows a reasonable correlation between the two. This is most likely due to a consistent approach being used to determine the over-surface for all the exchangers, and contradicts a position held by many engineers who prefer to use fouling factors. In this regression, the coefficient of determination r^2=0.853, implying a fair amount of variance, but the fit remains within reasonable range for a heat-transfer correlation. Based on this, the rough equivalence between fouling factor and over-surface is:

Table 4 Fouling Factor and Over-surface

	Fouling Factor (Total)	Over-surface
	ft^2-°F-hr/BTU (m^2-°C/w)	Percent
Normal exchangers	0.0075 (0.0013)	72
Critical service	0.0150 (0.0026)	145

These values are meant specifically for the design approach of keeping velocity constant. The approach is the one favored by most proponents of over-surface.

It is important to note how large the over-surface areas are. Even the commonly used fouling factors of 0.004 to 0.006 ft^2-°F-hr/BTU (0.0007-0.0011 m^2-°C/W) translate into over-surfaces of 38% to 58%. The recommended values in Table 4 are much higher.

8. Implications of Demonstrated Fouling Levels

The demonstrated fouling levels far exceed the commonly used TEMA values. The higher values in TEMA are approximately equal to the average fouling levels shown. The implications are:

- Most preheat trains are under-surfaced for the heat recovery expected.
- Little evidence exists to support optimization that uses different fouling factors on each side of the heat exchanger.
- Overall preheat train performance would meet design criteria for most units if the higher TEMA values are used; TEMA temperature, velocity, and split between crude fouling versus preheat stream fouling is ignored; and a modest allowance is made for uncertainty.
- Most energy optimization studies underestimate the surface area required, leading to an underestimate of the incremental cost of energy recovery. Estimates of energy efficiency improvements in most crude units overestimate the economic optimum energy recovery point.

Table 5 summarizes the incremental duty recovered versus the incremental area required for designing to different fouling factors in the units. This is based on the fouling factor required to get the same performance as a clean unit. Table 5 supports three important observations:

- The units have a relatively low return of incremental duty versus incremental area. Again, this supports that the benefits for incremental heat recovery mainly come from capacity and, potentially, quality.
- The incremental energy recovery for these units falls within a small range.
- The calculated result for incremental duty recovered per incremental area added is nearly the same whether using the actual fouling factor or the TEMA fouling factor. With a properly optimized unit, the return on investment for using the proposed (larger) fouling values for design is roughly the same as or better than the return for using the TEMA values. This leads to a series of extremely interesting investment decisions beyond the current scope. The section of Future Work expands on this observation.

Table 5 Incremental Heat Recovery versus Surface Area

	Unit 1	Unit 2	Unit 3	Unit 4
If designed to	% incremental duty / % incremental area			
TEMA Fouling Factor	0.12	0.11	0.17	0.09
Actual Fouling Factor	0.11	0.11	0.20	0.11

In addition, the results in Table 5 raise questions regarding optimizing preheat trains in their entirety versus optimizing single exchangers. The analysis is based on only four units, so the support of the optimization questions is not as strong as the conclusions for overall fouling factors[b]. This would be another useful area for further work.

9. Future Work

The present analysis comes to conclusions that are in conflict with some commonly held positions on exchanger design and preheat train design. It has also opened some questions that merit further in-depth study. These include:

- What are the benefits of improving preheat train performance with chemical additives (reducing fouling rates), alternative exchangers types (reducing fouling rates), and cleaning techniques (reducing cost of cleaning)? Can these shift the overall average fouling level before cleaning?
- Are preheat trains being designed for the correct fouling levels? Based on the near-constant incremental benefit in surface area versus heat recovery regardless of fouling levels, is the industry using the correct economic analysis for preheat train energy recovery? While incremental energy recovery and the associated capacity improvements have a linear benefit, the incremental cost for surface area typically follows a less-than-linear scale. This suggests that if using the TEMA fouling recommendations can be justified, then it can be equally justified to use the higher fouling factors suggested here. Inversely, if the higher values here are not justified, then using the TEMA values is not justified either. The current analysis supports a decision to move away from TEMA recommended values—but from this work it is not clear, in which direction.
- At what point do pump costs and the hydraulics of extra surface area affect economic decisions on which fouling factors or over-surface to use?
- Is the industry using the best techniques to monitor exchangers and decide when to clean them? Is poor decision-making the cause of the observed fouling levels at cleaning?
- What can be learned from more work on comparing fouling factor versus over-surface approaches to preheat train design?

10. Conclusions

Analysis of fouling factors experienced in normally operating crude units was extended from previous work [1]. In the original study, fouling rates from 49 exchangers were used and total fouling was extrapolated by assuming a conventional 180-day to 250-day period between cleanings. The number of units involved in this dataset was unclear, but it was at least two or three. New data examined in the current analysis include fouling factors from 105 exchangers in four units. These

[b] Overall fouling factor conclusions are based on this work plus a reasonable sample of exchangers from additional plants.

units were all running in a roughly steady-state of cleaning-fouling cycles. Combining the data creates a total dataset of 154 exchangers from multiple units.

The average fouling factor between cleanings was nearly identical for the two sets of data. The data support a total fouling design value of 0.0075 ft^2-°F-hr/BTU (0.0013 m^2-°C/W). This is suitable for overall design of crude trains. Problem services, and exchangers in critical services, should use a value of approximately 0.0150-0.0200 ft^2-°F-hr/BTU (0.0026-0.035 m^2-°C/W). Critical services include those where low duty would lead to downstream problems on the non-crude side, and exchangers in the last one or two services in the crude preheat train.

Fouling factor data show little evidence to support lower design fouling values for any of the following scenarios: specific services, raw crude versus desalted crude, and temperature dependence of the fouling factor before cleaning.

It is clear that desalting and temperature affect fouling rates. The total fouling before cleaning depends upon both the fouling rate and the duration of fouling. The decision to clean an exchanger is based upon the expected gain from a cleaner exchanger versus the total cost of cleaning the exchanger (cleaning cost plus lost duty during the cleaning). The data show that the vast majority of exchanger cleaning occurs when the fouling factor reaches 0.0150 ft^2-°F-hr/BTU (0.0026 m^2-°C/W). This gives a time-averaged fouling factor of 0.0075 ft^2-°F-hr/BTU (0.0013 m^2-°C/W).

Desalting crude oil is critically important. Desalting increases unit capacity, reduces downstream corrosion problems, improves energy efficiency in the fired heater, and helps preheat train operation. For the preheat train, it is important because it reduces the fouling rate and lengthens the interval between exchanger cleaning. Desalter performance is vital in units that must shut down for exchanger cleaning.

Temperature does affect fouling, but there is no clear correlation. For some crude blends the worst fouling occurred when the crude was coldest. For other crude blends the worst fouling occurred when the crude was hottest or in the middle temperature range.

Joshi et al [5] claimed a significant correlation between shear stress (velocity) and fouling rates. The exchangers reviewed in this work neither support nor contradict that analysis. Nearly all the new data are from exchangers with high crude velocities (>8 ft/sec, 2.43 m/sec) so the data set has too narrow a range to support conclusions about velocity versus fouling rates.

Over-surface values were determined with a constant-velocity approach to exchanger optimization. Contrary to commonly held thinking, over-surface showed a reasonable correlation with fouling factors. Equivalent over-surface factors were determined as 72% for normal design and 145% for critical services. These over-surface values are much higher than often used.

Because they use too low a fouling factor or too low an over-surface, most energy optimization studies underestimate the surface area required. This means that the incremental cost used for anticipated energy recovery is too low. Most energy efficiency improvement projects in crude units overestimate the economic optimum energy recovery point.

The overestimates become especially costly with aggressively energy-integrated units. High-temperature energy recovery requires high metal temperatures. Higher temperatures require expensive materials of construction to avoid corrosion. If this costly surface does not capture the predicted energy, expensive mistakes can be made during design and revamp projects. Many revamp projects have failed to earn the return that justified them. Refiner projects are nearly always capital constrained. Using the correct basis makes sure that capital is used most efficiently. Understanding fouling is the key to correctly using capital in new crude units and crude unit revamps.

11. Figures

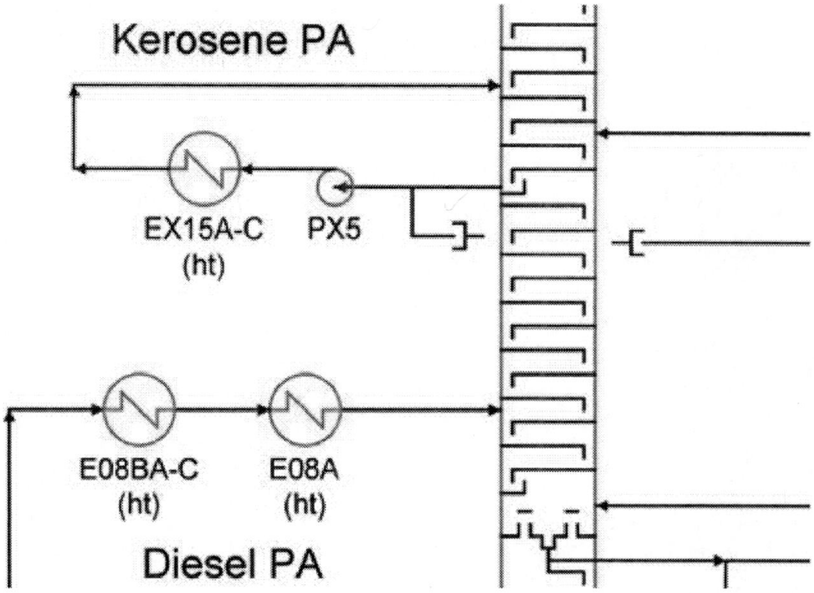

Figure 1 Preheat Train Exchangers Attached to Crude Distillation Tower

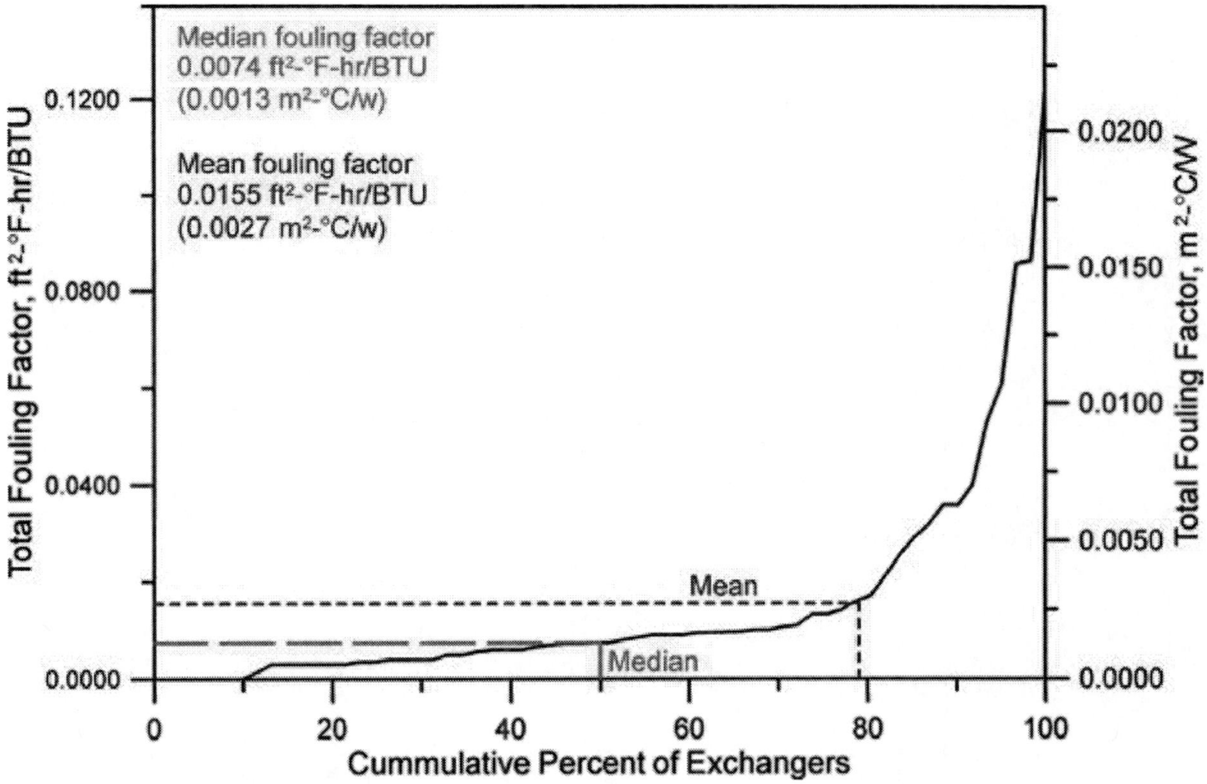

Figure 2 Fouling Factor Distribution

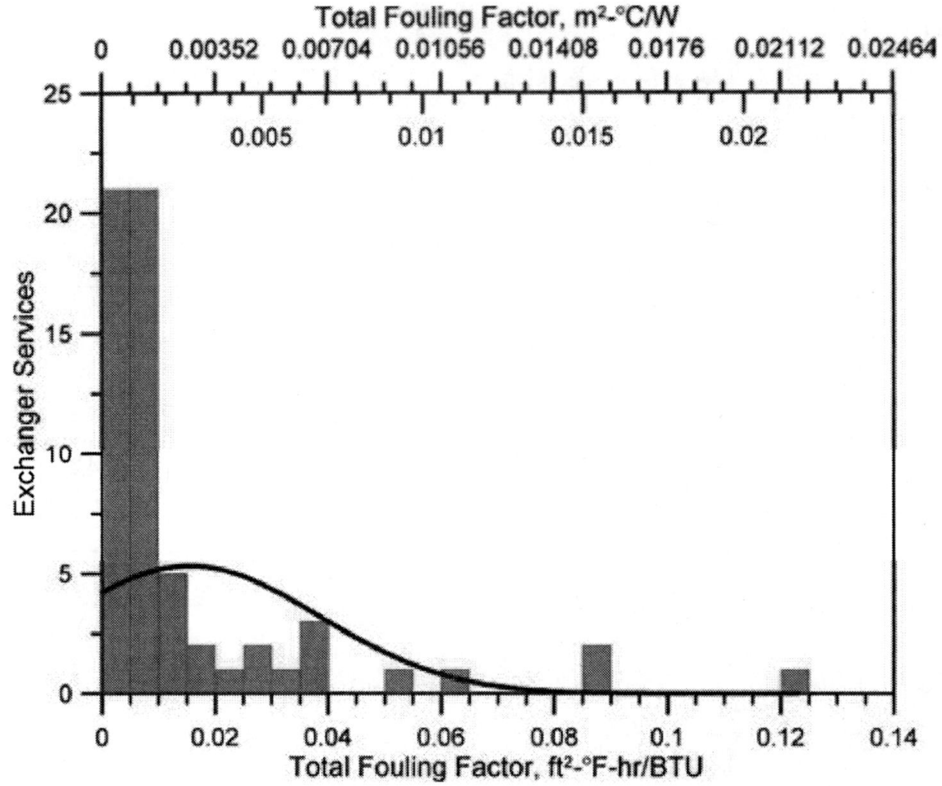

Figure 3 Exchanger Services versus Fouling Factor

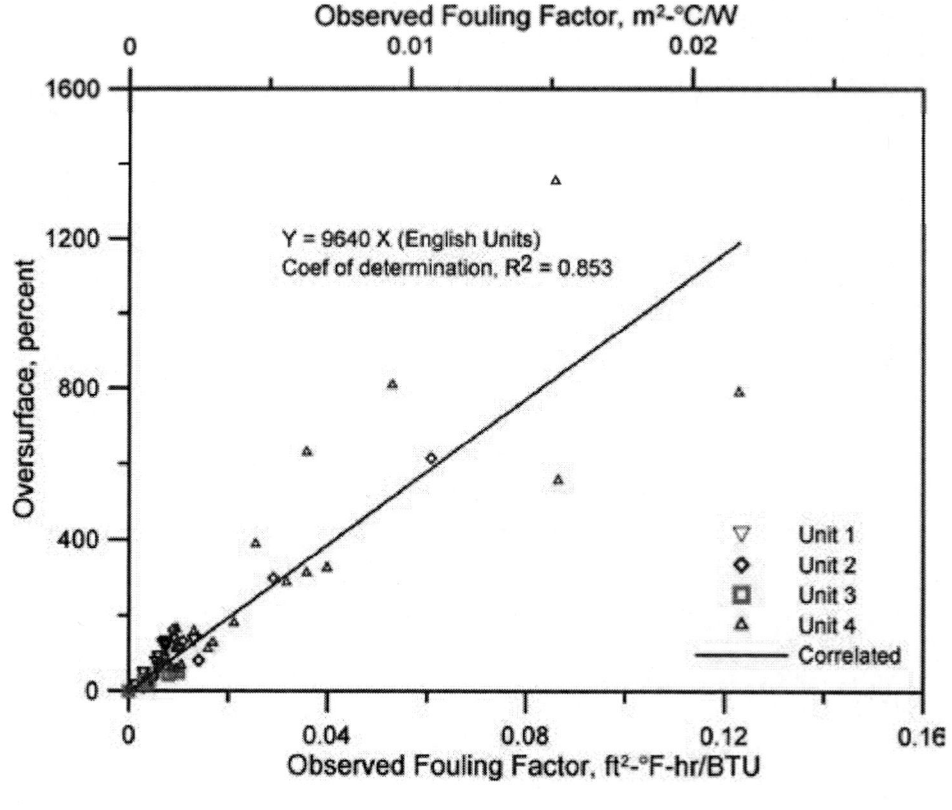

Figure 4 Over-surface versus Fouling Factor

12. References

[1] A. W. Sloley, "Crude Unit Fouling: What's Normal? Why's it Important?," presented at the AIChE Southwest Process Technology Conference, Galveston, TX, 2018.

[2] Anon., "Standards of the Tubular Exchanger Manufacturers Association, Ninth Edition," Tarrytown, NY, Standard, 2007.

[3] J. M. Chenoweth, "Final Report of the HTRI/TEMA Joint Committee to Review the Fouling Section of the TEMA Standards," *Heat Trans Eng*, vol. 11, no. 1, pp. 73–106, 1990.

[4] Various, "Proceedings Heat Exchanger and Cleaning Conferences (multiple)." Heat Exchanger Fouling and Cleaning, Various [Online]. Available: http://www.heatexchanger-fouling.com/proceedings.htm. [Accessed: 19-Feb-2019]

[5] H. M. Joshi, N. B. Shilpi, and A. Agarwal, "Relate Crude Oil Fouling Research to Field Fouling Observations," in *Proceedings of the International Conference on Heat Exchanger Fouling and Cleaning VIII*, Schladming, Austria, 2009 [Online]. Available: http://www.heatexchanger-fouling.com/papers/papers2009/2_Joshi_MS.pdf

549362 *Accelerating Product Innovation at Dow through Multivariate Modeling*

Accelerating Product Innovation at Dow through Multivariate Modeling

In the era of Industry 4.0, manufacturers such as Dow are focused on gaining more value from their historical data than ever before. ProSensus has distinguished itself as a trusted, global leader in Big Data analytics by helping many Fortune 500 companies across numerous industries use their data to innovate for the future.

ProSensus has recently worked with Dow to accelerate innovation on one of their key product lines – silicone antifoams. For product development applications such as this, ProSensus combines powerful multi-block latent variable modeling with constrained optimization. This approach allows clients to simultaneously optimize the selection of raw materials, recipe formulations, manufacturing conditions and costs to reach targeted product performance properties while adhering to custom constraints. A model-based approach allows Dow to rapidly develop and scale up custom silicone antifoam formulations for the dynamic technical and regulatory needs of customers in the pulp and paper, food and beverage, wastewater treatment, metal working, and other industries.

This presentation will examine the approach taken (data assembly, multivariate modeling, model validation, and optimization), challenges encountered, and results obtained. Perspectives from both ProSensus and Dow will be included.

550548 Mechanical Reliability in the Chemical Industry: Challenges and Successes in Predictive Maintenance Modeling

To be successful in the chemical industry, safety is first and foremost and reliability is key. Unexpected equipment failures can mean lost production, higher emissions from abnormal conditions and process swings, and more risk to those working in the facility. Over the past decades, maintenance strategies have evolved in the chemical industry to minimize unexpected equipment failures along with the associated cost and production losses. Historically this is done by understanding mechanical failure mechanisms, creating preventative maintenance programs based on the mean time between failures, and monitoring key performance indicators.

There exists a wealth of data in the chemical industry in process instrumentation and historical equipment records, and advances in digitization and data utilization are occurring in the age of Industry 4.0. Data mining, increased computing power, and user-friendly implementations of machine learning models enable early detection of impending failures, in some cases with significantly improved lead times compared to condition based monitoring systems. However, challenges exist in modeling these sparse datasets (few failures to train a model on), on equipment that is not like-for-like across an enterprise, and where the end-users do not have data science expertise.

In this talk we will discuss the improvements in reliability that machine/deep learning models can provide along with the challenges the chemical industry faces in applying these methodologies. Dow Chemical is undergoing an evaluation of many current technologies on the market, and we will share key elements in our assessment of the capabilities and the fit for our industry. There are challenges with the "big data" we have available, but there are successes to share as well, ultimately leading to more reliable, sustainable, and safe chemical plants.

550311 Deep Learning-Based Sensor for Crystal Size and Contour Characterization

Deep Learning-based Sensor for Crystal Size and Contour Characterization

Vidhyadhar Manee, Wenbo Zhu, José A. Romagnoli

Department of Chemical Engineering, Louisiana State University, Baton Rouge, LA, 70809

Abstract

In the field of crystallization, practitioners do not have access to sensors that can make real-time, accurate measurements of crystal size in high-density slurries. Several paths have been explored to solve this problem but most of them have come up short when applied to industrial settings. One path that has shown some promise is video microscopy and image analysis. In this approach, a probe is inserted into the crystal slurry and a microscope captures images. The images are sent to an imaging software, where the individual crystals are segmented using variations of a two-step algorithm: morphological treatment followed by particle segmentation. This approach works well when the particles are adequately separated but breaks down in dense slurries. Particles in dense slurries have their shapes distorted by attrition and experience significant overlap due to surface forces. This obscures the intra-particle boundaries and complicates the task of segmenting crystal particles. Detecting particles in such an environment requires a sensor that is robust to large variations in particle size, shape and visibility.

In recent years, the field of computer vision has switched over from traditional feature extraction-based algorithms to more powerful tools like deep learning (DL) and artificial intelligence (AI). This transition has led to impressive breakthroughs in a variety of computer vision challenges like image labelling, object detection and instance segmentation. The success of these tools has spurred researchers to revisit unresolved industrial problems and discover new solutions.

In this work, we propose a deep learning-based sensor that solves the problem of crystal detection in high-density slurries using the Mask RCNN, a state-of-the-art CV model. The sensor functions in three stages. In the first stage, it analyzes the image pixels and extracts meaningful features like edges, textures, patterns, parts and objects. In the second stage, it uses this information to weed out the background and identify locations in the image that enclose crystal particles. In the final stage, the coordinates of the locations are iteratively refined to tightly enclose the particles and segmentation masks are generated to isolate the contours of individual particles. This information is then used to precisely quantify the shape characteristics of the crystals and generate the size distribution.

The Mask RCNN architecture used in the sensor is powered by a ResNet-101 backbone and incorporates approximately 63 million parameters. All the parameters were pre-trained on the COCO dataset and one-third of them were further fine-tuned on the crystal dataset. Our dataset has 100 labeled training images and 20 labeled validation images, each with at least 70 crystal particles. The original images were resized to 1024x1024 pixels and extensive data augmentation was used to prevent overfitting. The model converges in approximately 30 epochs, takes 4 hours to train and can process images up to a speed of 5 FPS on a GPU. It shows promising results in segmenting crystal particles in a wide variety of conditions. Figure 1 shows the results for the anti-solvent crystallization of sodium chloride in water using ethanol as anti-solvent in an experimental bench-scale semi-batch crystallizer.

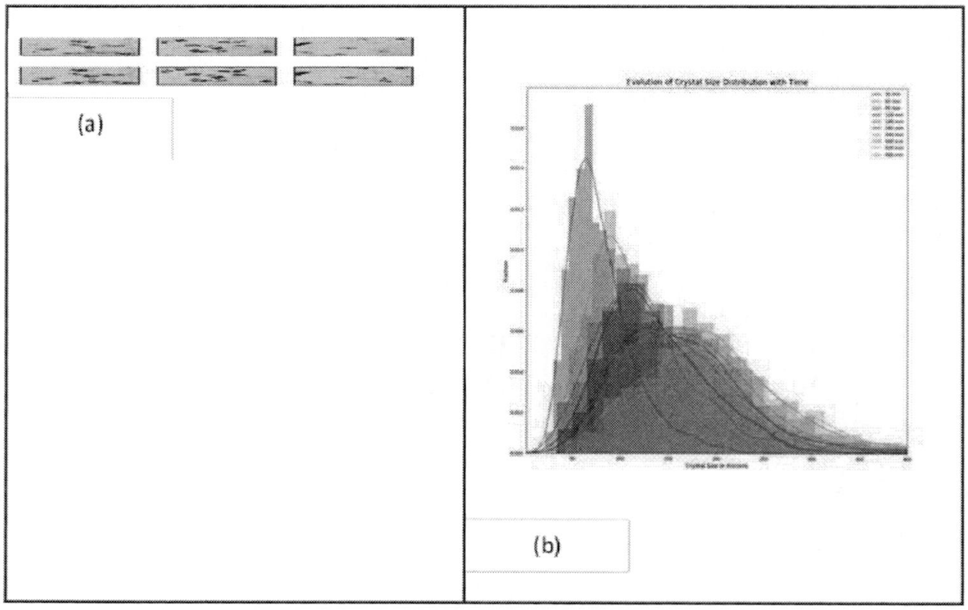

Figure 1. (a) Crystal segmentation results for each crystal instance. (b) The crystal size distribution drawn from the segmentation results.

551241 Reinforcement Learning (RL) Based Data-Driven Decision Support System

Big data has transformed from a buzzword to a real value creator. In recent years there has been a rising interest in data analytics in the energy industry. The industry and academia have made significant progress in big data analytics. However, there is still a need to use the plant data to gain insights and address the issues related to abnormal situation management. An abnormal situation is defined as "a disturbance or series of disturbances in a process that causes plant operations to deviate from their normal operating state." The disturbance can be a minor near-miss to a catastrophic incident which may cause production losses, environmental damage, injuries or sometimes fatalities if not managed properly. During an abnormal situation the operator is expected to take appropriate action to bring back the process to a normal state. Diagnose and mitigation actions depend on the operator's experience, process complexity and stress levels of the operator.

This paper outlines a systematic framework for application of Reinforcement Learning (RL) during process operations and abnormal situation management. A novel method using Reinforcement Learning is established to address decision making challenges and advise operators during abnormal situations. The system is designed on the expert knowledge from operators and engineers, the knowledge available from previous operation experiences and process plant information. The application of the proposed method can be expanded by connecting the system to Operator Training Simulators (OTS) and train operators for abnormal situation management.

A SEQUENCE TO SEQUENCE TIME SERIES FORECASTING MODEL FOR BATCH PROCESSES WITH SEASONALITY

Shu Xu, Mark Nixon

Emerson Automation Solutions

Introduction

Modern chemical manufacturing processes generally fall into three categories: continuous, batch and semi-batch. Unlike its continuous counterpart where the raw materials are constantly fed into the system and processed into an intermediate or final product continuously, a batch process receives and processes the feedstock by a series of discrete steps. Such a manufacturing process has been widely used in semiconductor, pharmaceutical, specialty chemical and other industries. For example, the fermentation process used in brewing industry. Because the whole process consisting of many operating steps can hardly be operated in a steady state, it poses tougher challenges for both equipment and operators to assure product quality in many batch runs. One specific challenge is related to seasonality which stems from equipment aging[1], periodic maintenance[2], catalyst decay [3] and so on. Typical examples of such a behavior involve activity coefficients of reactor catalysts and fouling factors in heat exchangers[4]. Process engineers are challenged to forecast such a seasonality lurking in the system so that they can become proactive in identifying and preventing potential problems.

In this paper, we built a sequence-to-sequence model to forecast the seasonality simulated in the benchmark Penicillin bio-reactor process. The results obtained by such a sequence-to-sequence model are compared with those generated from a linear partial least square (PLS) method and a nonlinear support vector machine (SVM) regression method.

The remainder of this paper is organized as follows. Section 2 briefly introduces methodology information regarding time series forecasting, long short-memory (LSTMs) networks, and an associated sequence-to-sequence model. Section 3 provides case study details including the benchmark Penicillin bio-reactor simulator and simulation conditions. Section 4 summaries results obtained in the case study. We conclude the paper and discuss future works in Section 5.

Methods

Time series forecasting

During batch operations, abundant informative data can be collected in the form of time series with regular sampling periods [5]. As a result, forecasting methods in time series analysis, which take advantage of dependence between observations in the time dimension to predict the future, can be applied in process seasonality forecast [6], for example, classical methods include autoregression, autoregressive integrated moving average (ARIMA), simple exponential smoothing (SES), etc. In addition, time series forecasting can also be reframed as a supervised learning problem. By reframing the time series data via a lag method, various linear and nonlinear supervised learning algorithms can be used, including linear regression, support vector machine, neural networks, or a sequence-to-sequence model with long short-term memory (LSTM) networks as discussed in this paper.

LSTM network

Long short-term memory (LSTM) networks belong to the category of recurrent neural networks (RNNs, see Figure 1) where loops and chains exist to allow information to persist. In Figure 1, an input x_t is fed into a regular neural network block A which produces an output h_t. An RNN distinguishes itself from its regular counterpart in that each neural network unit in the RNN can pass information either to itself in a loop or to a successor to form a chain. Although a normal RNN can track the short-term dependence between observations in the time series, it struggles to learn the "long-term" connections [7]. To overcome such a challenge, the LSTM network is proposed[8] (see Figure 2).

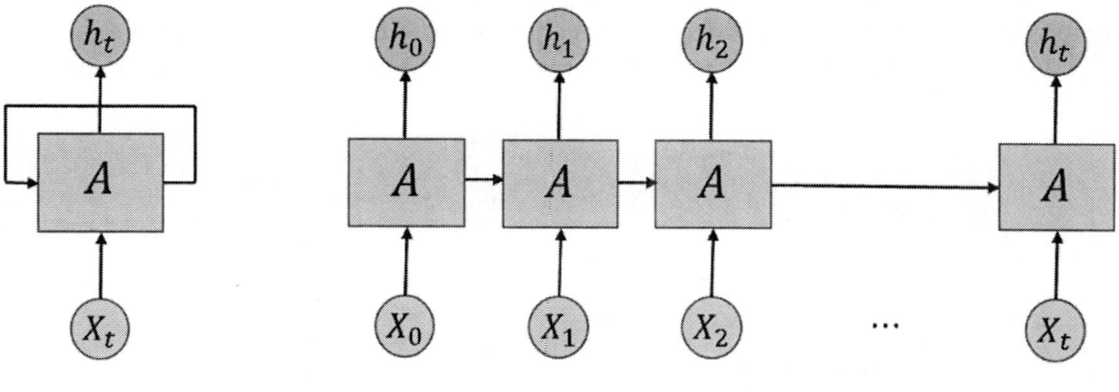

(a) An RNN with loop (b) An unrolled RNN

Figure 1 Recurrent neural network demonstration

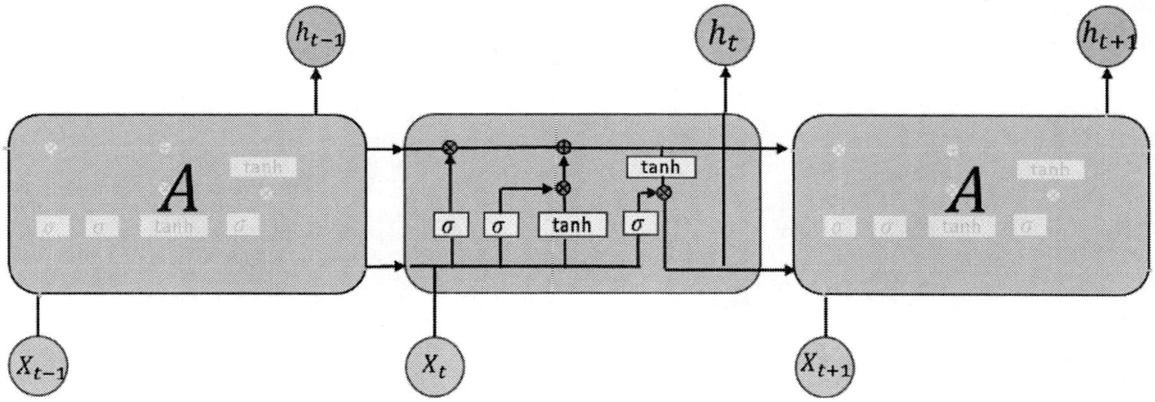

Figure 2 LSTM network demonstration

Although the general form of the LSTM network is akin to that of a regular RNN, its repeating cell each contains four neural network layers interacted in a way designed specifically for remembering information for long periods of time. A detailed description of the structure of each cell is shown as follows.

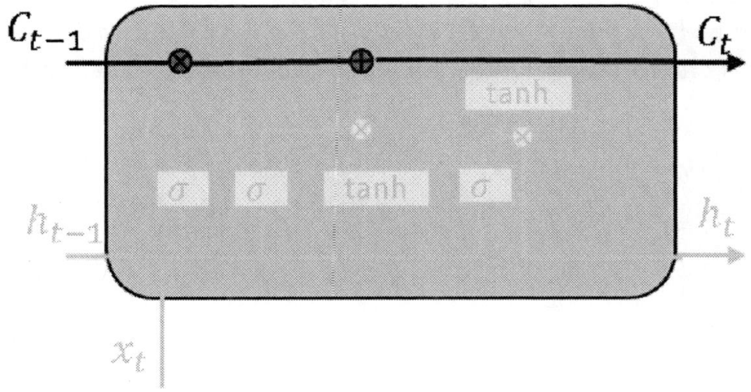

Figure 3 Cell state in an LSTM cell

As illustrated in Figure 3, a horizontal line running across the top of cell along with only linear pointwise operation indicates potentially a free flow of information without any cell status changes. In addition, the amount of cell status information gain or loss is controlled by the following gates in Figure 4, which are comprised of a sigmoid neural net layer and a pointwise multiplication operation. Because the sigmoid layer outputs a number between 0 and 1, it regulates how much information should be let through. While a value of 0 indicates closing the gate, a value of 1 means letting the information flow through freely.

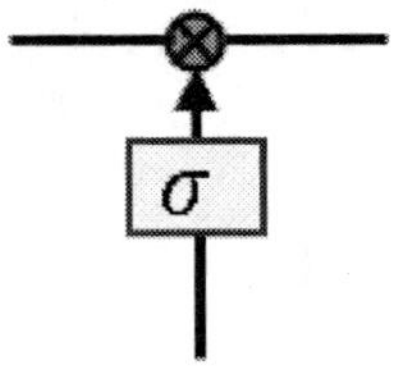

Figure 4 Gate in an LSTM cell

Figures 5~8 provide a step-by-step walk-through of the four-layer structure inside each LSTM cell. As shown in Figure 5, the first layer named the "forget gate layer" decides whether to keep the previous cell state C_{t-1} or not. For example, if the forgetting factor f_t calculated by gate Equation (1) equals 0, the previous cell state is completed discarded (or "forgotten") after an element-wise multiplication.

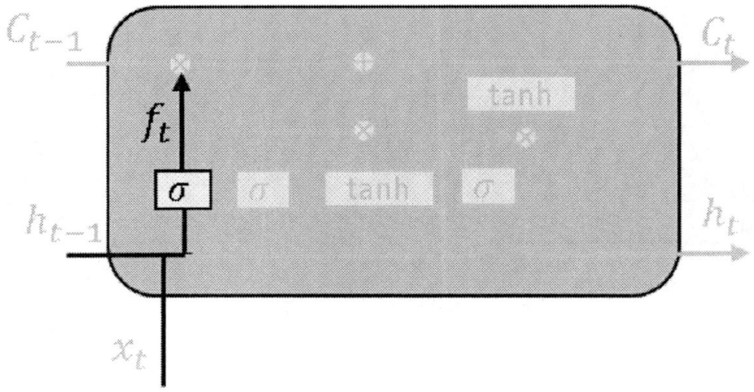

Figure 5 Forget gate layer in an LSTM cell

$$f_t = \sigma\left(W_f \cdot [h_{t-1}, x_t] + b_f\right) \tag{1}$$

The next state update layer consists of two sub-layers, out of which the "input gate layer" decides whether we need to update the cell state with the new value produced by the "new value layer", as shown in Figure 6. The associated gate and value equations are listed in Equations (2) and (3) respectively.

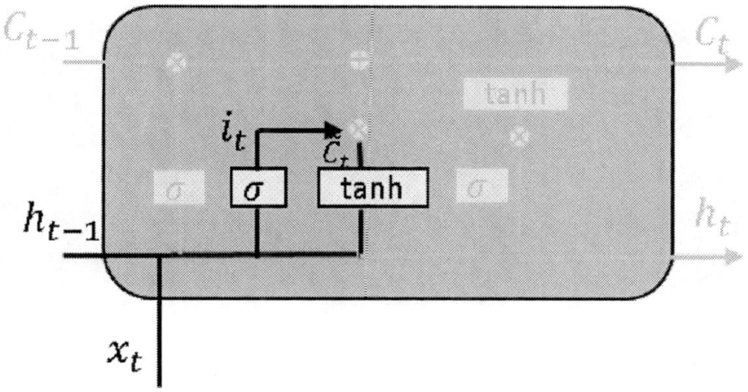

Figure 6 State update layer in an LSTM cell

$$i_t = \sigma(W_i \cdot [h_{t-1}, x_t] + b_i) \tag{2}$$

$$\tilde{C}_t = tanh(W_C \cdot [h_{t-1}, x_t] + b_C) \tag{3}$$

Now since we have both the previous cell state C_{t-1} and the new state update value $\tilde{C}_t$, the new cell state C_t can be calculated by Equation (4), and the diagram of state update is exhibited in Figure 7.

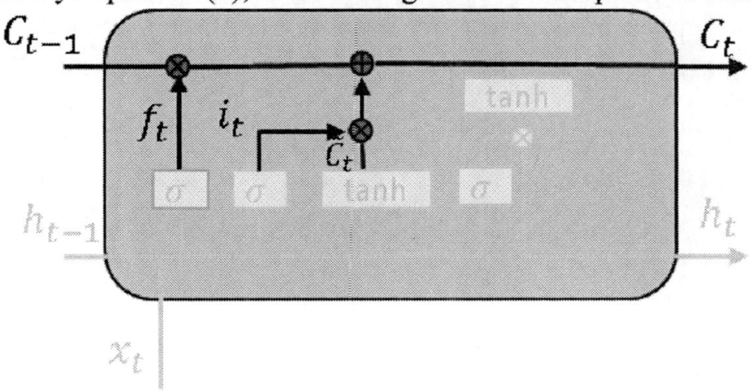

Figure 7 State update operation in an LSTM cell

As shown in Equation (4), the new state C_t equals the sum of an old state C_{t-1} multiplied by the forgetting factor f_t, and a new state update $\tilde{C}_t$ scaled by a factor i_t.

$$C_t = f_t * C_{t-1} + i_t * \tilde{C}_t \tag{4}$$

Finally, the output of the cell is decided by the output layer shown in Figure 8, including a sigmoid gate which decides whether we need to deliver a filtered version of the cell state C_t, the h_t, and associated equations are shown in Equations (5) and (6).

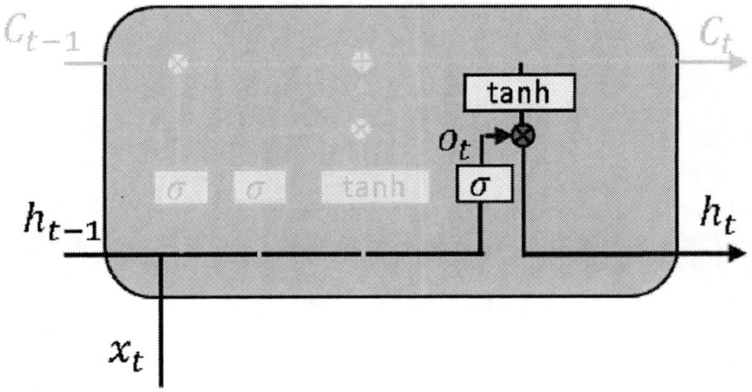

Figure 8 Output layer in an LSTM cell

$$o_t = \sigma(W_o \cdot [h_{t-1}, x_t] + b_o) \tag{5}$$

$$h_t = o_t * tanh(C_t) \tag{6}$$

In our study, a sequence to sequence (seq2seq) model with LSTM cells is used for time series forecasting. The model includes an encoder converting an input sequence X_t to corresponding state vectors which contain transformed context for each input point x_t, and a decoder which takes as input such context vectors, internal hidden states to generate an output sequence Y_t.

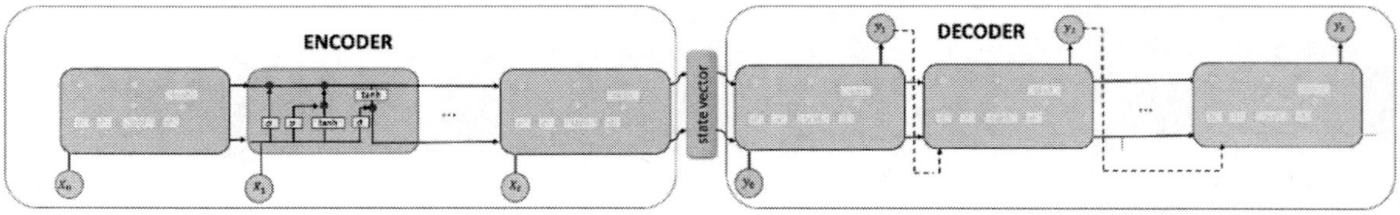

Figure 9 A seq2seq model with LSTM cells

Case study

The Benchmark Penicillin Bio-reactor simulator

A lumped-parameter dynamic batch-fed Penicillin model developed by Birol et al. [9] is used as the basis for current study. The original MATLAB code for the simulation program can be downloaded from Dr. Cinar's website (www.chee.iit.edu/~cinar/software.htm), and modifications have been made on the key kinetic parameter to simulate the batch seasonality. The main state variables included in this simulation are: reactor temperature, biomass volume, pH, biomass concentration, substrate concentration, penicillin concentration, dissolved oxygen concentration, carbon dioxide concentration. The full model is provided in the Appendix. The ODE23t stiff solver provided in MATLAB is used to simulate the process.

The general process flow diagram of the Penicillin reactor is shown in Figure 10. In the initial growth phase, the fermenter is initiated with a small volume of bacteria culture and substrate (glucose and oxygen). The reactor temperature is kept at 298 K (room temperature) to initiate the growth of the Penicillin producing bacteria. After approximately 50 hours, the initial substrate is depleted; the reactor then enters the penicillin production phase. In this phase, the fermenter is operated in semi-batch mode. Continuous flows of substrate (through the substrate tank) and oxygen (through the aerator in the bottom of the fermenter) promote the production of penicillin, carbon dioxide and other by-products.

Furthermore, two PID controllers regulate the acidity level (pH ~5) and temperature (298 K) of the fermenter to promote Penicillin production.

Figure 11 plots the resulting trajectory of a single batch. The sampling frequency is set to once every 30 minutes, mimicking that of real plant data. In addition, additional output variables of the simulation include agitator power, aeration rate, substrate flow rate, cooling water flow rate, heat generation. These variables are either manipulated inputs, or measured variables of the PID control loop.

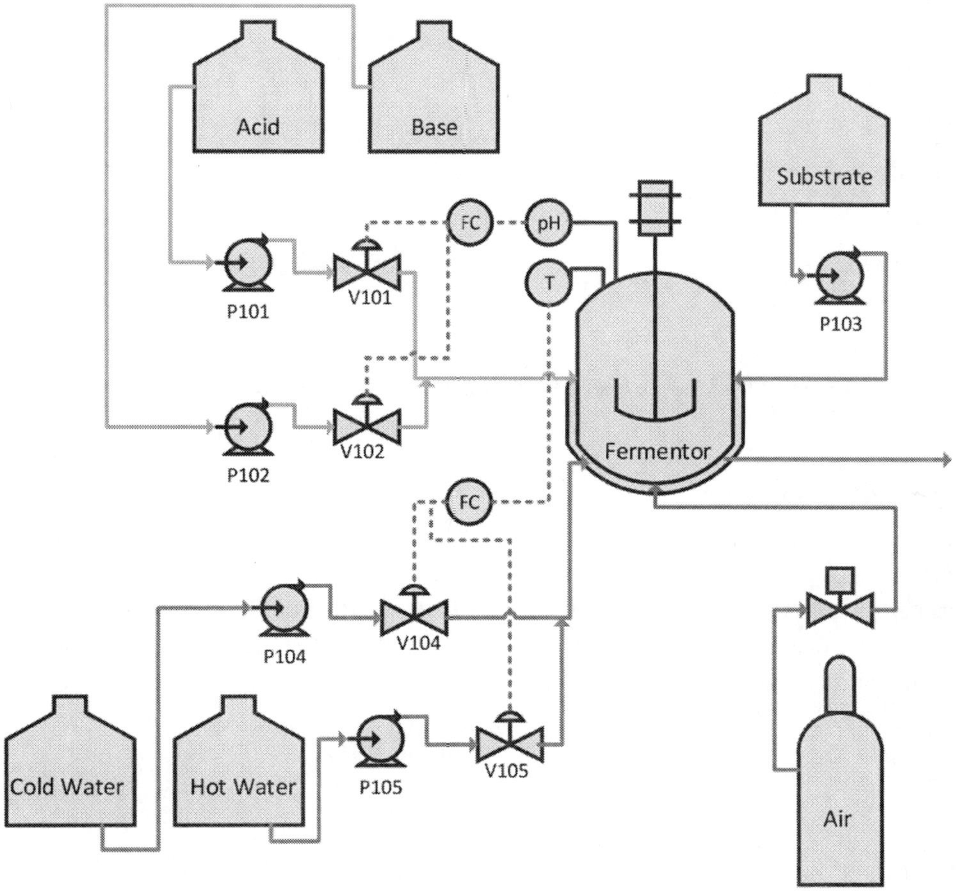

Figure 10 Process flow diagram of the batch-fed penicillin process

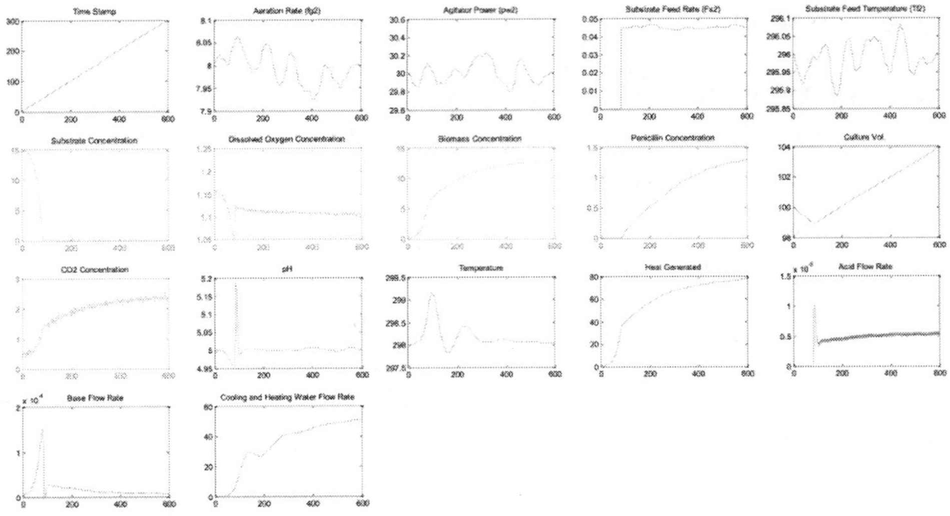

Figure 11 Example batch trajectories from a simulation run of the penicillin reactor

Simulation Conditions

To introduce process seasonality, the specific growth rate μ in the biomass growth equation (see Equation (7)) is configured to follow a first-order exponential decay with time constant set to 150 batches. The decay of the specific growth rate μ starts from the nominal value of 0.092 till the terminal value of 0.048. As shown in Figure 12, scheduled maintenance (occurs once every 100 batch cycles) of the reactor will reset the specific growth rate back to its nominal value.

$$\frac{dX}{dt} = \mu X - \frac{X}{V}\frac{dV}{dt} \quad (7)$$

Table 1 lists the simulation conditions for our current study. A total of 600 batches with seasonality corresponding to six cycles in Figure 12 are simulated, out of which 400 batches are used for training, and the remaining 200 are used for testing. All the data goes through batch-wise unfolding as shown in Figure 13, where 3D batch data are flattened batch-wise into a 2D matrix. After unfolding, data from the ith batch are squeezed into the ith row of the matrix, and such a row is comprised of values of variable j measured at time k. After scaling and mean centering of the flattened matrix, a lag of 10 batches are used in matrix reformatting for supervised model building. In other words, we use prior 10 batches of data to forecast the next batch.

All the time series forecasting models, including the sequence-to-sequence (seq2seq), support vector machine (SVM) regression, and partial least squares (PLS), are built in python, where libraries such as tensor flow, sklearn, pandas, and numpy are used.

Table 1 Simulation conditions for batch data generation

Case	Training	Testing
	400 batches with seasonality	200 batches with seasonality

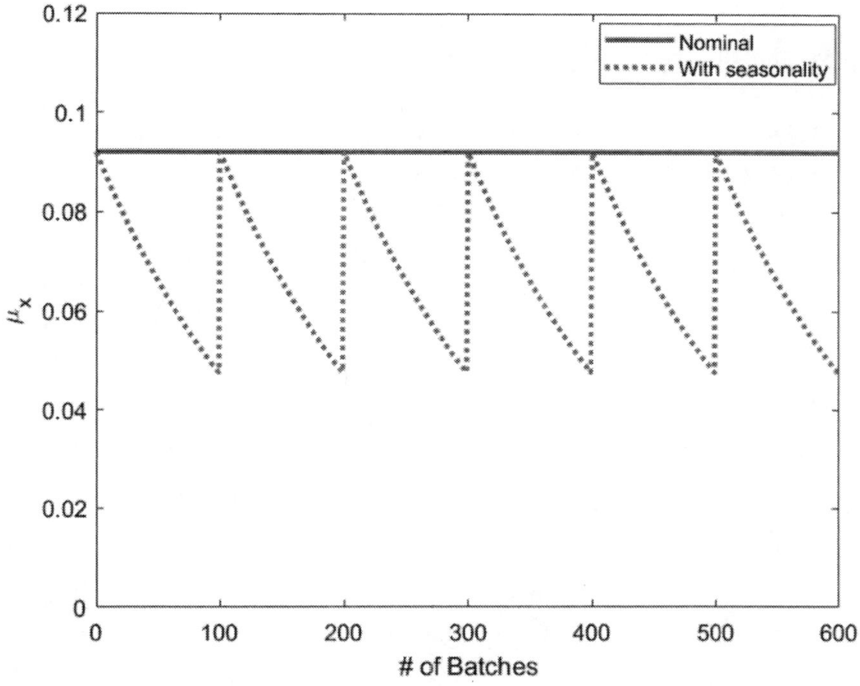

Figure 12 Seasonality in specific growth rate of bacteria culture over six cycles (decay and reset)

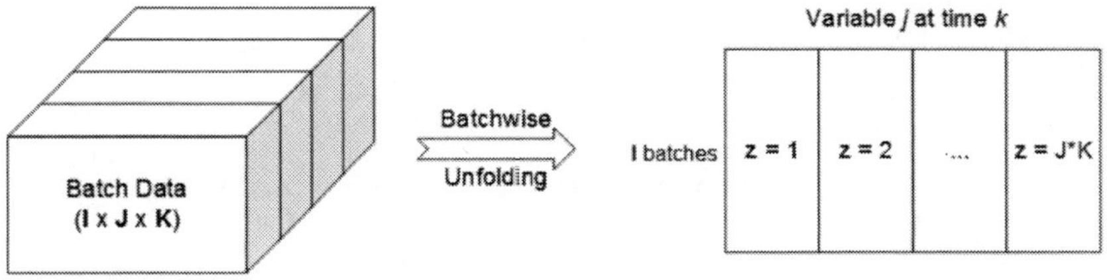

Figure 13 Batch-wise unfolding scheme in time series forecasting

Results and Discussions

In this study, we use the root mean square error of forecasting calculated in Equation (8) to evaluate the forecasting performance, and the associated results are listed in Table 2. It is observed that the seq2seq model outperforms the other two supervised learning methods in terms of forecast accuracy of batch seasonality.

$$RMSEF = \sqrt{\frac{\sum_{i=1}^{n}(y_i - \hat{y}_i)^2}{n}} \quad (8)$$

Table 2 RMSEF values of forecasting models

Forecasting models	RMSEF
Seq2seq	3.41

SVM	4.40
PLS	4.46

The actual forecasting results are illustrated in Figures 14~16. It is observed that the nonlinear models including the seq2seq and SVM obtain smoother forecasting results than the linear PLS model. In addition, the seq2seq model excels in forecasting the exact batch where the production of penicillin rises abruptly due to reset of bacteria growth rate.

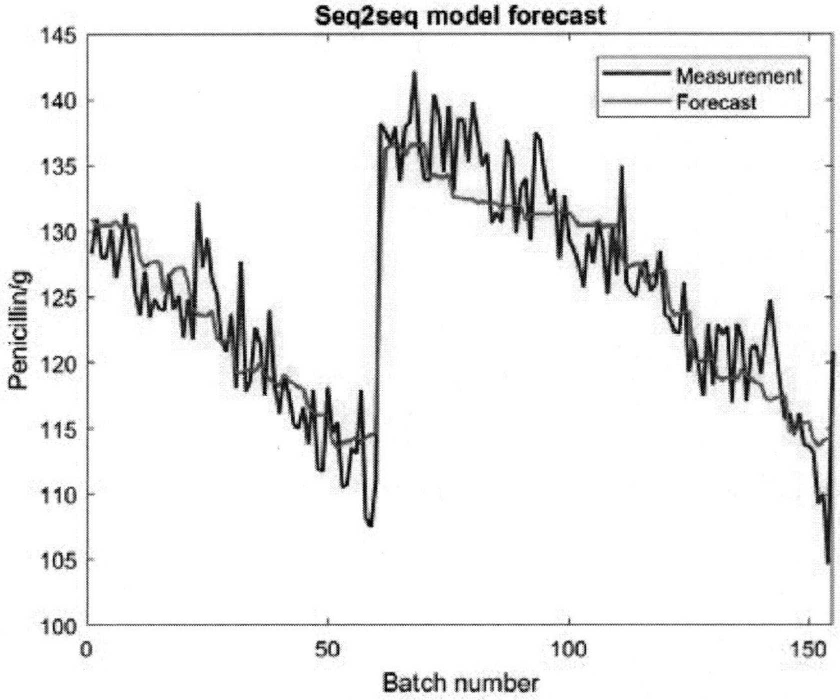

Figure 14 Seq2seq model forecasting results

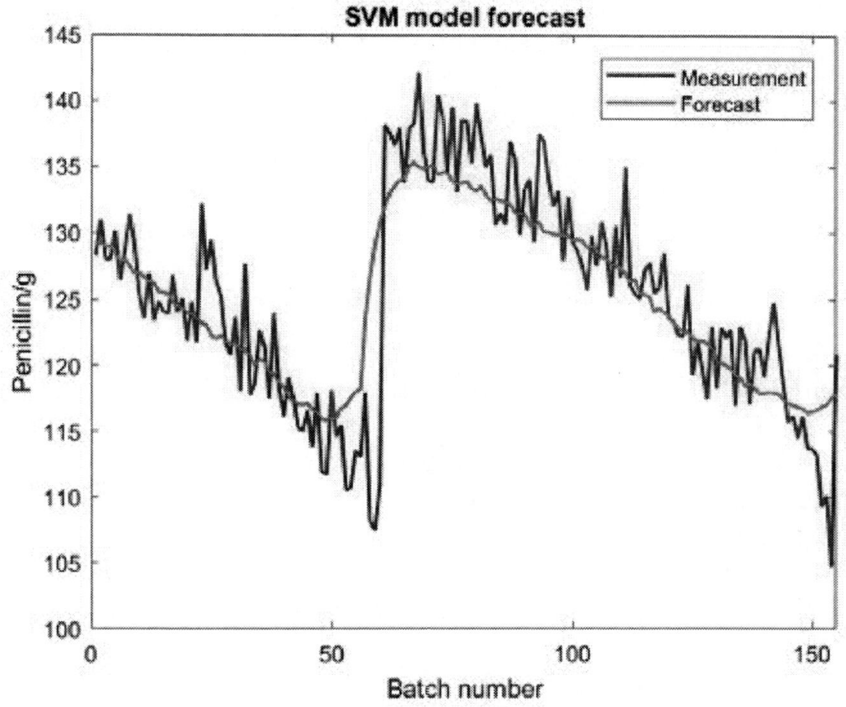

Figure 15 SVM model forecasting results

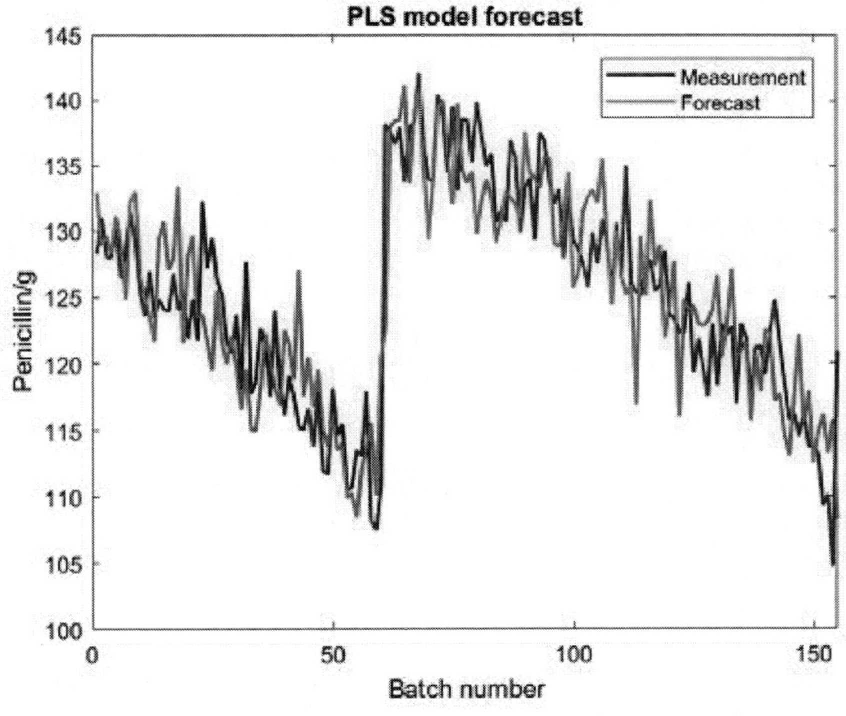

Figure 16 PLS model forecasting results

Conclusions and future work

In current work, we demonstrate the adoption of the sequence to sequence learning framework to overcome the challenge of batch seasonality forecast. A time series forecasting model is built in Tensor

Flow, and it is tested on the data generated from the benchmark Penicillin bio-reactor simulator. The model successfully forecasts the cyclic change of penicillin production rate with the periodic decay of the biomass growth rate. Such results illustrate the successful application of data analytics to wrestle with real problems in chemical manufacturing within the Digital Transformation era.

In the future, we will test such a seq2seq framework on more industrial data and compare its performance with other time series forecasting methods.

References

[1] C. Clarotti, A. Lannoy, S. Odin, and H. Procaccia, "Detection of equipment aging and determination of the efficiency of a corrective measure," *Reliab. Eng. Syst. Saf.*, vol. 84, no. 1, pp. 57–64, Apr. 2004.

[2] M. Ji, Y. He, and T. C. E. Cheng, "Single-machine scheduling with periodic maintenance to minimize makespan," *Comput. Oper. Res.*, vol. 34, no. 6, pp. 1764–1770, Jun. 2007.

[3] B. W. Wojciechowski, "A theoretical treatment of catalyst decay," *Can. J. Chem. Eng.*, vol. 46, no. 1, pp. 48–52, 1968.

[4] H. Müller-Steinhagen, *Heat exchanger fouling: mitigation and cleaning techniques*. IChemE, 2000.

[5] S. J. Qin, "Process Data Analytics in the Era of Big Data," *AIChE J.*, vol. 60, no. 9, pp. 3092–3100, 2014.

[6] G. E. P. Box, G. M. Jenkins, G. C. Reinsel, and G. M. Ljung, *Time series analysis: forecasting and control*. John Wiley & Sons, 2015.

[7] Y. Bengio, P. Simard, P. Frasconi, and others, "Learning long-term dependencies with gradient descent is difficult," *IEEE Trans. neural networks*, vol. 5, no. 2, pp. 157–166, 1994.

[8] S. Hochreiter and J. Schmidhuber, "Long short-term memory," *Neural Comput.*, vol. 9, no. 8, pp. 1735–1780, 1997.

[9] G. Birol, C. Ündey, S. J. Parulekar, and A. Çınar, "A morphologically structured model for penicillin production," *Biotechnol. Bioeng.*, vol. 77, no. 5, pp. 538–552, 2002.

[10] G. Birol, C. Ündey, and A. Cinar, "A modular simulation package for fed-batch fermentation: penicillin production," *Comput. Chem. Eng.*, vol. 26, no. 11, pp. 1553–1565, Nov. 2002.

Appendix – Penicillin Batch Reactor Model

Modeling Equations

The modeling equations and associated parameters, constants are taken from Birol et al. [10] and are simply provided here for completeness and convenience.

Biomass Growth

$$\frac{dX}{dt} = mX - \frac{X}{V}\frac{dV}{dt}$$

where μ is the specific growth rate and contains the effects of environmental variables (pH and Temperature) as well as carbon source (glucose) and oxygen in its kinetic expression.

Penicillin Production:

$$\frac{dP}{dt} = m_{pp}X - KP - \frac{P}{V}\frac{dV}{dt}$$

where μ_{pp} is the specific penicillin production rate containing biomass, carbon source (glucose) and oxygen concentrations in its kinetic expression. Here, the hydrolysis of penicillin is also considered and expressed as a first order rate expression with a rate constant K.

Substrate Utilization:

Glucose:

$$\frac{dS}{dt} = - \frac{m}{Y_{xs}}X - \frac{m_{pp}}{Y_{ps}}X - m_X X + F - \frac{S}{V}\frac{dV}{dt}$$

Oxygen:

$$\frac{dC_L}{dt} = - \frac{m}{Y_{xo}}X - \frac{m_{pp}}{Y_{po}}X - m_o X + K_{la}\left(C_L^* - C_L\right) - \frac{C_L}{V}\frac{dV}{dt}$$

Here, K_{la} is taken to be a function of agitator power input and flow rate of oxygen as suggested by Bailey and Ollis (1986).

Carbon Dioxide Production:

$$\frac{dCO_2}{dt} = a_1\frac{dX}{dt} + a_2 X + a_3$$

The rate expression for CO_2 has been taken from the earlier work of Montague et al. (1986).

Effect of pH

$$\frac{d\left[H^+\right]}{dt} = \gamma\left(\mu X - \frac{FX}{V}\right) + \left[\frac{-B + \sqrt{\left(B^2 + 4\times10^{-14}\right)}}{2} - \left[H^+\right]\right]\frac{1}{\Delta t}$$

Where $B = \dfrac{\left[10^{-14}/\left[H^+\right] - \left[H^+\right]\right]V - C_{a/b}\left(F_a + F_b\right)\Delta t}{V + \left(F_a + F_b\right)\Delta t}$, F_a and F_b represents acid and base flow rates in l/h, respectively.

Volume Change

$$\frac{dV}{dt} = F + F_{a/b} - F_{loss}$$

where $F_{a/b}$ is the flow of acid/base, F_{loss} accounts for the evaporative losses during fermentation:

$$F_{loss} = V\lambda\left(e^{5(T-T_o)/(T_v-T_o)} - 1\right)$$

Heat of Reaction

$$\frac{dQ_{rxn}}{dt} = r_{q1}\frac{dX}{dt}V + r_{q2}XV$$

where r are yield coefficient associated with the exothermic microbial reactions.

Energy Balance

$$\frac{dT}{dt} = \frac{F}{s_f}\left(T_f - T\right) + \frac{1}{V\rho c_p}\times\left[Q_{rxn} - \frac{aF_c^{b+1}}{F_c + \left(aF_c^b / 2\rho_c c_{pc}\right)}\right]$$

PID controller

A discretized PID control law has been used in both the temperature and pH controls. E_N denotes the error at time step N and E_{N-1} denotes the error at time step N-1.

$$\Delta MV_N = K_c\left(E_N - E_{N-1} + \frac{\Delta t}{\tau_1}E_N - \frac{\tau_d}{\Delta t}\times\left(CV_N - 2CV_{N-1} + CV_{N-2}\right)\right)MV_N$$

Overall, there are total of 15 differential equations which are solved simultaneously. All the parameters are taken from literature or assigned values which are physically meaningful in the absence of available literature. Keeping in mind that most of the parameters are strain specific and also are affected by the nature of the substrate and the environmental conditions like pH and temperature, the user should be aware of the fact that the parameters used in this simulator are average values rather than specific to a certain strain or an environmental condition.

Initial Conditions, Kinetic and Controller Parameters

Initial conditions	
Substrate concentration: S (g/l)	15
Dissolved oxygen concentration: C_L ($= C_L^*$ at saturation) (g/l)	1.16
Biomass concentration: X (g/l)	0.1
Penicillin concentration: P (g/l)	0
Culture volume: V (l)	100
Carbon dioxide concentration: CO_2 (mmol/l)	0.5
Hydrogen ion concentration: $[H^+]$ (mol/l)	$10^{-5.1}$
Temperature: T (K)	297
Heat generation: Q_{rxn} (cal)	0
Kinetic parameters and variables	
Feed substrate concentration: s_f (g/l)	600
Feed flow rate of substrate: F (l/h)	
Feed temperature of substrate: T_f (K)	298
Yield constant: $Y_{x/s}$ (g biomass/g glucose)	0.45
Yield constant: $Y_{x/o}$ (g biomass/g oxygen)	0.04
Yield constant: $Y_{p/s}$ (g penicillin/g glucose)	0.90
Yield constant: $Y_{p/o}$ (g penicillin/g oxygen)	0.20
Constant: K_1 (mol/l)	10^{-10}
Constant: K_2 (mol/l)	7×10^{-5}
Maintenance coefficient on substrate: m_x (per h)	0.014
Maintenance coefficient on oxygen: m_o (per h)	0.467
Constant relating CO_2 to growth: α_1 (mmol CO_2/g biomass)	0.143
Constant relating CO_2 to maintenance energy: α_2 (mmol CO_2/g biomass h)	4×10^{-7}
Constant relating CO_2 to penicillin production α_3 (mmol CO_2/l h)	10^{-4}
Maximum specific growth rate: μ_x (per h)	0.092

Contois saturation constant: K_x (g/l)	0.15
Oxygen limitation constant: K_{ox}, K_{op} (no limitation)	0
Oxygen limitation constant: K_{ox}, K_{op} (with limitation)	2×10^{-2}, 5×10^{-4}
Specific rate of penicillin production: μ_p (per h)	0.005
Inhibition constant: K_p (g/l)	0.0002
Inhibition constant for product formation: K_I (g/l)	0.10
Constant: p	3
Penicillin hydrolysis rate constant: K (per h)	0.04
Arrhenius constant for growth: k_g	7×10^3
Activation energy for growth: E_g (cal/mol)	5100
Arrhenius constant for cell death: k_d	10^{33}
Activation energy for cell death: E_d (cal/mol)	50 000
Density × heat capacity of medium: ρ C_p (per l °C)	1/1500
Density × heat capacity of cooling liquid: ρ_c C_{pc} (per l °C)	1/2000
Yield of heat generation: r_{q_1} (cal/g biomass)	60
Constant in heat generation: r_{q_2} (cal/g biomass h)	1.6783×10^{-4}
Heat transfer coefficient of cooling/heating liquid: a (cal/h °C)	1000
Cooling water flow rate: F_c (l/h)	
Constant: b	0.60
Constants in K_{la}: α, β	70, 0.4
Constant in F_{loss}: λ (per h)	2.5×10^{-4}
Proportionality constant: γ (mol $[H^+]$/g biomass)	10^{-5}

Controller parameters (PID)	
pH: (base) K_c, τ_I: (h), τ_d: (h)	8×10^{-4}, 4.2, 0.2625
pH: (acid) K_c, τ_I: (h), τ_d: (h)	1×10^{-4}, 8.4, 0.125
Temperature: (cooling) K_c, τ_I: (h), τ_d: (h)	70, 0.5, 1.6
Temperature: (heating) K_c, τ_I: (h), τ_d: (h)	5, 0.8, 0.05

Progress Toward Machine Learning Method for Corrosion Surveillance of Nuclear Materials Storage Containers

Bruce Hardy[1], Jason Bakos[2], Taylor Clingenpeel[2], Anna dEntremont[1], Michael Martinez-Rodriguez[1], Brenda Garcia-Diaz[1] and Lindsay Roy[1],
1. Savannah River National Laboratory, Aiken, SC,
2. Computer Science and Engineering, University of South Carolina, Columbia, SC

Introduction

Within the DOE complex, 3013 canisters are used to store Pu-bearing waste. Wall penetration due to corrosion, specifically stress corrosion cracking, is considered to be the most likely cause of failure over the 50-year lifetime of the canisters.

The 3013 canister system consists of three nested canisters: an innermost "convenience can", containing the Pu-bearing material, that is sealed inside an inner 3013 canister serving as the primary containment, which is, in turn, placed in an outer 3013 canister that serves as a secondary containment, see Figure 1. The safety basis for the 3013 system, mandates that there is no gas leakage from the inner 3013 can. The top of the inner 3013 canister is sealed by welding, and the weld-affected region is called the Inner Can Closure Weld Region (ICCWR), see Figure 2. Over time, halides contained in the Pu-bearing material can be released from the convenience can into the gas space within the inner canister. In simulated exposures using test coupons, exposure to the halides have been found to produce corrosion preferentially in the ICCWR.

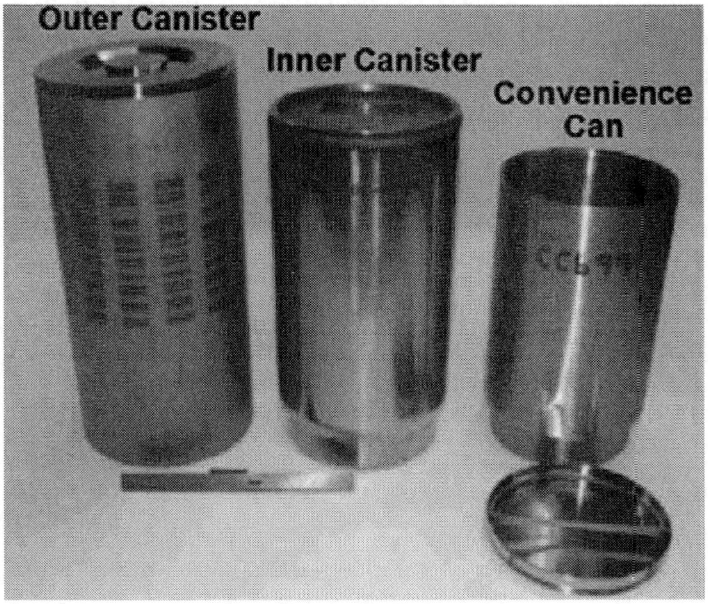

Figure 1 3013 nested canister system.

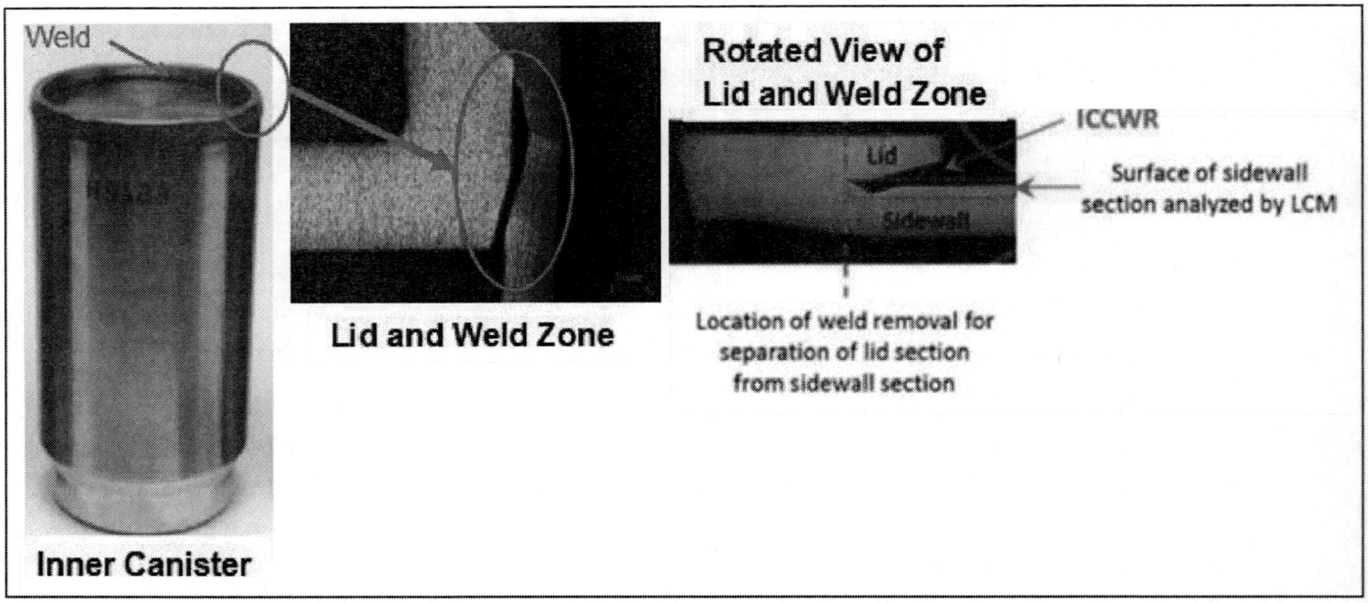

Figure 2 Inner Can Closure Weld Region (ICCWR) and sectioned canister.

Corrosion inspection is part of the Material Identification and Surveillance (MIS) program for long term storage of plutonium-bearing materials in the DOE complex. It is an objective of the MIS program to develop evaluation methodologies and metrics for assessment of corrosion. Currently, the principal method for corrosion inspection is image data obtained using a Laser Confocal Microscope (LCM). In the inspection process, randomly selected canisters are cut into sections and examined via LCM imaging. The LCM produces binary data files that contain surface height data, RGB color data (both optical and [optical + laser intensity]) and laser intensity data, yielding 8 layers of data per pixel. There are approximately 6000 images produced per can with each image measuring 1024 by 768 pixels, yielding 786,432 pixels per image, with the aforementioned 8 layers of data for each pixel. The scans are used to analyze statistics for features associated with corrosion. After establishing the statistical protocol, SRNL must work through a 5-year backlog of images while continuing to receive new can images for analysis. Initially, manual inspection was attempted, but given the vast quantity of data and its complexity, an effort was undertaken to implement numerical methods including computer vision and machine learning algorithms.

Processing of LCM Data

Corrosion is strongly, but not exclusively, associated with surface pitting and cracking. Coloration, shapes and patterns of surface features are also indications of corrosion. Conversely, not all pits and surface lesions are the result of corrosion, some are artifacts of fabrication, impact, scoring or other non-corrosion events. In manual observations, corrosion is identified via the combined properties of pit depth, area, edge contour, color and clustering. Examples of progressively degraded surfaces are shown in Figure 3. Surface data, representing physical characteristics, are extracted from the binary LCM files and used to identify features of interest. Surface characteristics, indicative of corrosion, are obtained by applying mathematical operations to the features. The mathematical operations are those invoked for computer vision and include, but are not limited to, gradients, correlations and filters.

In the most simplistic application, image data is used directly by flagging parameters, such as pit depth and cracking, that are associated with the incipience or presence of corrosion. Flagged surface features are then examined manually for final determination of whether corrosion is has occurred. While this approach greatly improves the efficiency of the examination process, it is still be very time consuming. A more sophisticated approach is to assess the data using machine learning algorithms to identify corrosion without manual intervention. The latter is the goal of the work discussed in this paper.

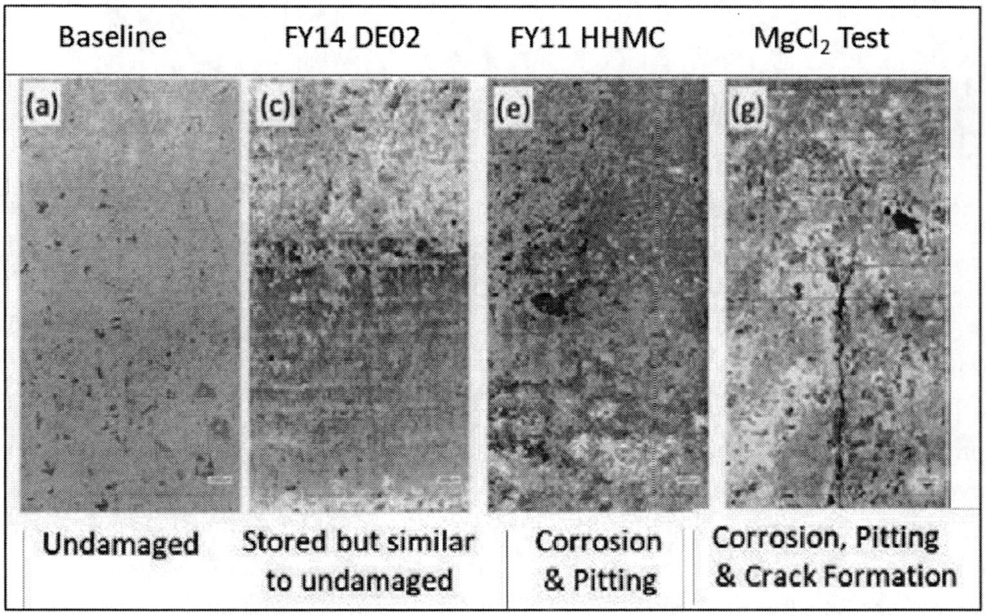

Figure 3 Optical images ranging from undamaged to severely damaged ICCWR. The HHMC is a canister exposed to high humidity and the $MgCl_2$ test was a sample exposed to boiling $MgCl_2$ to accelerate corrosion.

It goes without saying that any approach to numerical evaluation requires the binary LCM data to be converted back to physical properties; namely height, RGB color and grayscale image values. Recovery of the physical properties can be accomplished using software such as Matlab®[1] or Gwyddion[2]. However, the LCM data requires additional processing to make it useful for identification of corrosion. Explaining this requires a description of the way that a sample is examined by the LCM. For example, consider the computation of the local surface height, which is a measure of the depth of pits, or the height of protrusions, relative to the neighboring surface. In the LCM measurement process, the inner canister is first sectioned and placed into a sample holder, see Figure 4. The samples have curvature, due to the cylindrical shape of the canister, and tilt, due to the way they are mounted in the holder. LCM height measurements provide the absolute surface height, which is the combined result of tilt, curvature, pits cracks, etc. Clearly, for corrosion associated features, such as pits, the height data of interest is taken relative to the neighboring surface. However, the height of the surface neighboring the feature is the average of the local surface height, which is a function of tilt and curvature effects. A GUI was developed using Matlab® App Designer to simplify the input and processing of the LCM data. Through the GUI, the user is able to: generate plots, images and histograms; flag surface structures exceeding input thresholds; and select either single images or, select the number of images to be stitched into a mosaic, for interrogation.

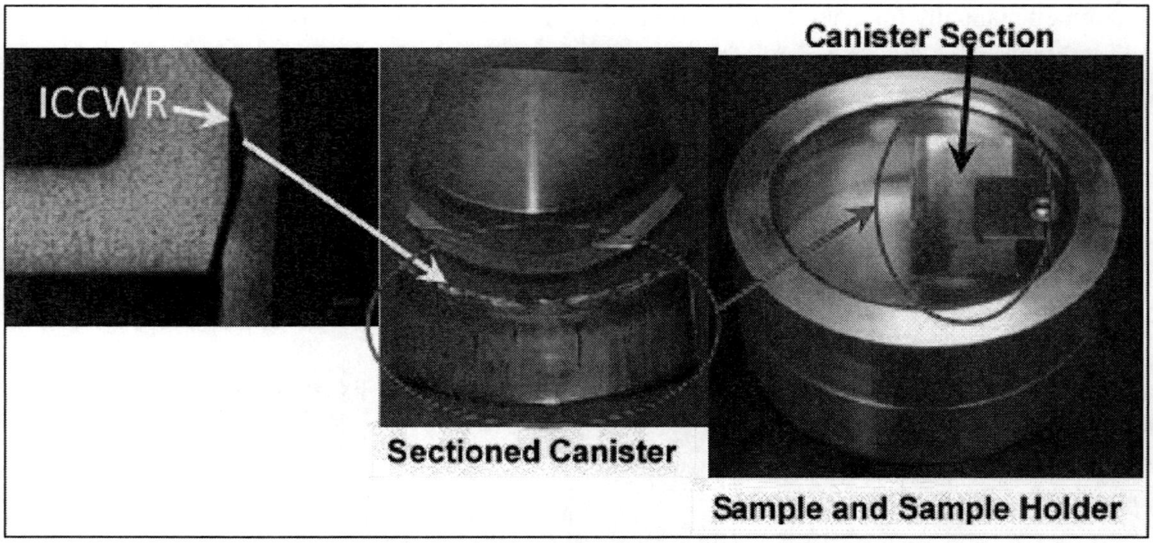

Figure 4 Sectioned inner canister and loading of section in sample holder

To obtain the general physical surface features, LCM data is extracted from large binary files, with software written to convert the data to physical attributes (i.e. height, color and grayscale values; all as functions of location on a projection of the canister surface onto a plane). Since the sample surface is not flat, the software computes the mean surface height by applying the user's choice of either a polynomial regression of varying degree or a spectral method (SMOOTHN[3]) The mean surface height is then used as a reference to determine the depth of depressions and protrusions. The user interface for the software permits selective downloading of the binary data and interrogation of the attributes. User input thresholds are used to flag attributes of interest. Physical attributes are used to compute features associated with corrosion, such as the volumes, shape and cross-sectional areas of depressions, pits, cracks and prominences. The features are accessible to permit further interrogation of the surface data through the user interface, as well as for use in machine learning algorithms.

The LCM output data consists of a collection of images, each of which is a very small part of the overall sample surface area. Individual images are rectangular, composed of 1024 by 768 pixels, with each pixel being a square that is 694.92nm long on each side. To obtain a large-scale perspective of the surface, the images must be stitched together to form a mosaic. A mosaic composed of stitched images is important when dealing with features that span many images, permitting a more complete overview of surface structure. However, a difficulty arises due to the way the LCM outputs height data on an image-wise basis. For each image, the LCM measures surface height relative to an arbitrary reference. This means that the absolute heights reported by the LCM can differ discontinuously between adjacent images, resulting in a vertical mismatch between the images along the stitched edges, producing results such as those shown in Figure 5. This figure shows the absolute surface height for adjacent images taken for the ICCWR. The software corrects for this effect by using the average height difference between edges of adjacent images to match the reference surfaces, see Figure 6. The stitching routine accounts for variable overlap between adjacent images by matching the height of the reference surface at the edges of the individual LCM images. In tests with 3013 data the matching of height along the edges of stitched images results in very little apparent error. An array of 300 stitched images is shown in Figure 7. Here, the ridges along the top-right and bottom-left borders of the right-hand image representing height relative to the local surface result from the abrupt drop-off in height as the scan passes beyond the cut edges of the sample, which is too abrupt to be captured by the polynomial reference surface.

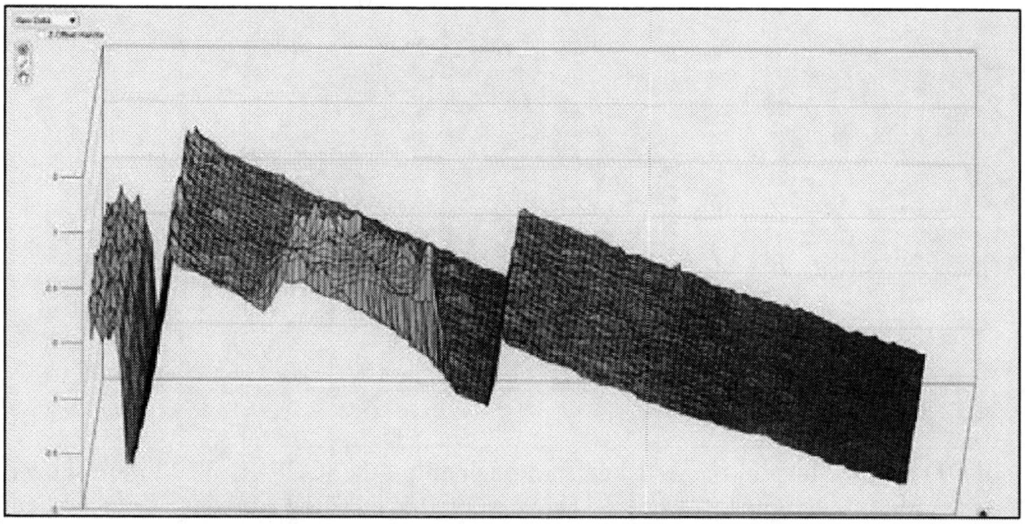

Figure 5 Vertical height mismatch between stitched images from uncorrected LCM data. Absolute surface height is shown. The x-y (area) scale is 694.92nm/unit, while the z (height) scale is 0.1nm/unit.

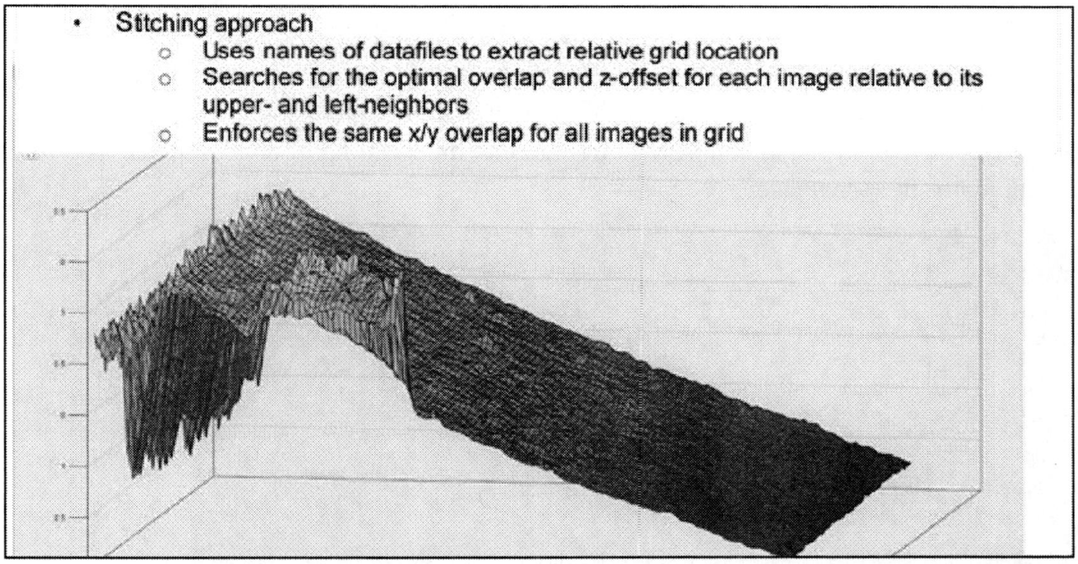

Figure 6 Correction of vertical height mismatch between stitched images using averaged height difference between edges of adjacent images. Absolute surface height is shown. The x-y (area) scale is 694.92nm/unit, while the z (height) scale is 0.1nm/unit.

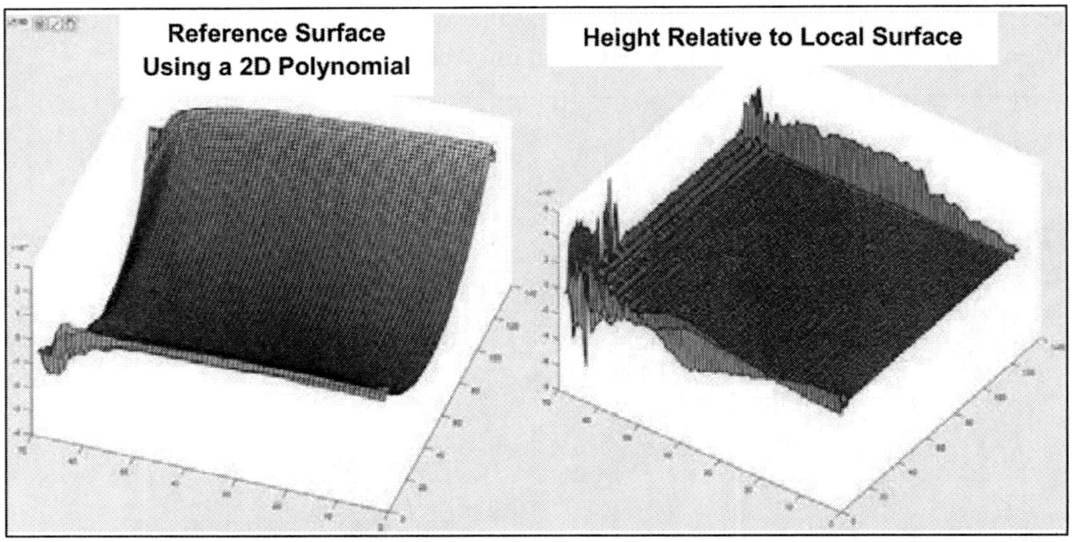

Figure 7 Array of 300 stitched height images taken for a segment of the ICCWR. The figure shows a 12.8mm x 6.9mm section, consisting of 18390 x 9980 pixels (183 Mpixels), with an x and y scale of 694.92nm for each side of a pixel. Note, without overlap the image would be 18432 x 9984 pixels.

For RGB data, assuming the same overlap as for the height data resulted in an aberration that spanned approximately 1-2 pixels. This suggests that it may be incorrect to assume that all LCM channels produce the same image-wise overlap. Possible explanations include an inherent spatial mismatch between different detectors or a distortion caused by taking LCM data through a viewport that is used to isolate radioactive canister sections. Even so, as can be seen in Figure 8, the stitching of optical images produces a mosaic that is continuous in appearance.

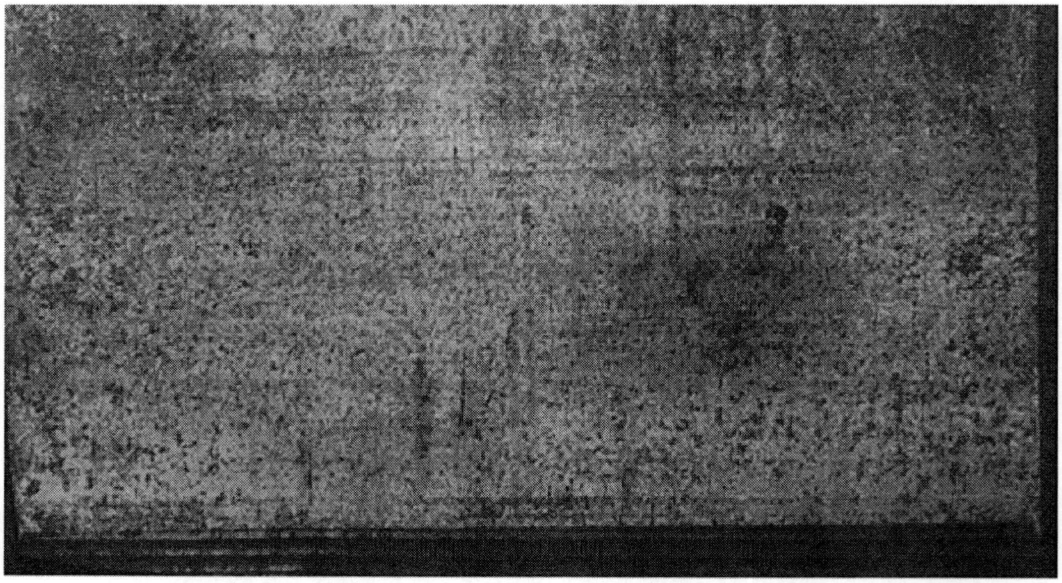

Figure 8 Array of 300 stitched optical images taken for a segment of the ICCWR. As for Figure 6, this figure represents an 12.8mm x 6.9mm section, consisting of 18390 x 9980 pixels (183 Mpixels), with an x and y scale of 694.92nm for each side of a pixel. Note, without overlap the image would be 18432 x 9984 pixels.

Results

In the current version of the image software, features are extracted using threshold-based rules that are applied after correction for sample tilt and curvature. Pits or protrusions of potential interest are identified by limits input by the user. Adjacent pixels that exceed the threshold are grouped into a single feature, which defines a region. Features flagged by the thresholds are compiled into a list for labeling and future processing. In addition to image plots and data values, the software can generate histograms of feature data, which will be important for statistical analysis in future stages of this work.

While pits and surface protrusions can be identified via height data, cracks, especially very narrow ones, require optical data for their detection. Optical features defining a crack are defined by edges. For this reason, edge detection algorithms were explored for application to RGB/optical features. One method for detection of edges in optical features is to look for large gradients in color, intensity or other picture functions. Alternative approaches include covariances of picture functions and filters. Methods considered for detection include differencing masks (Prewitt, Sobel and Roberts) and filters (Gaussian and LOG-Laplacian of Gaussian). Along with potential as edge detection methods, filters, which can be applied as masks, have the benefit of reducing image noise as well. Under certain conditions, it was found that covariances with RGB data were a bit better at edge detection than gradient techniques. However, the application of covariances relied on proper setting of thresholds. Overall, the superiority of either method was not definitively established.

Results from height and RGB covariance feature detection methods are shown in Figure 9. Figure 10 shows the difference in the size of a feature based on different LCM data. In this figure, the feature appears to have reduced area when expressed in terms of height data relative to that for optical data. The reason for this difference is that a sufficient depth is required to produce a distinct pattern in the plot of height data, while optical data relies on surface characteristics, such as roughness and color. In other words, the feature under consideration has shallow surface damage surrounding it that can be detected optically due to a difference in reflectivity (which may be wavelength dependent). The resulting optical image thus consists of the deep parts of the feature together with the surrounding shallow surface damage. However, the surface damage bounding the feature is sufficiently shallow that is does not produce much of a change in the color palette corresponding to height. Thus, the feature appears smaller when viewed in terms of height rather than optical data. The sensitivity of apparent feature size to the type of data has important implications for automated interrogation of the LCM images.

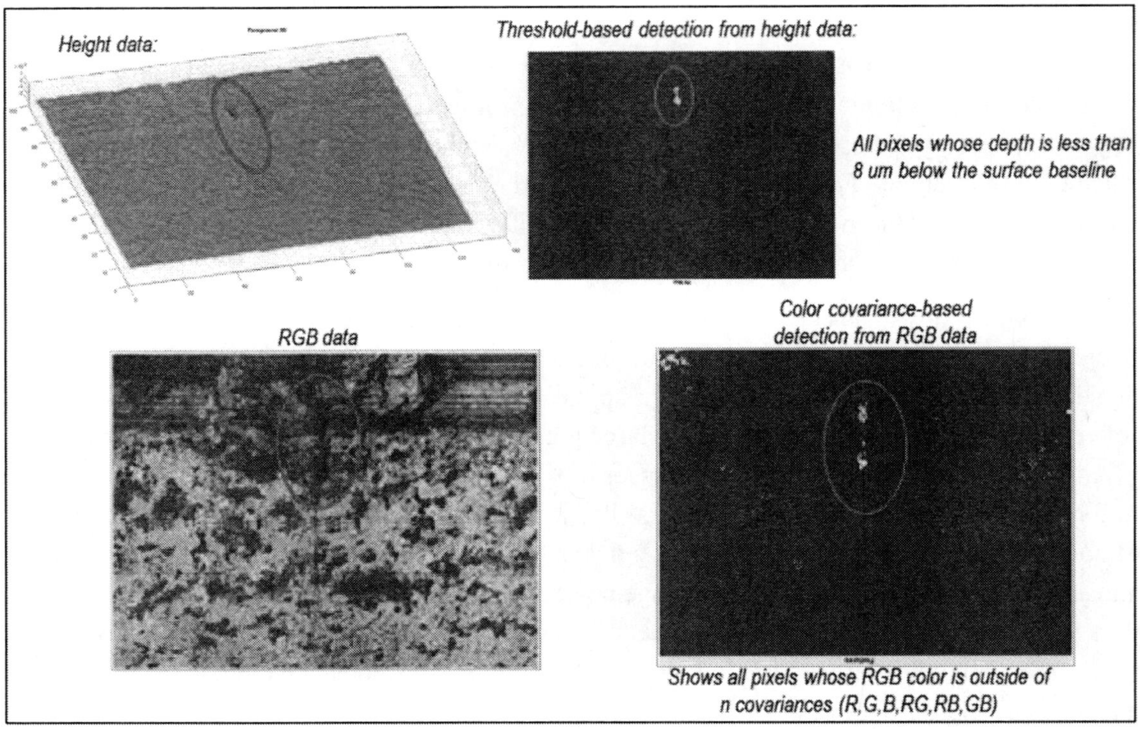

Figure 9 Feature detection via height threshold and RGB covariance threshold, shown in the upper and lower right figures, respectively. The upper left figure is local height relative to the reference surface. The lower left figure is RGB optical data.

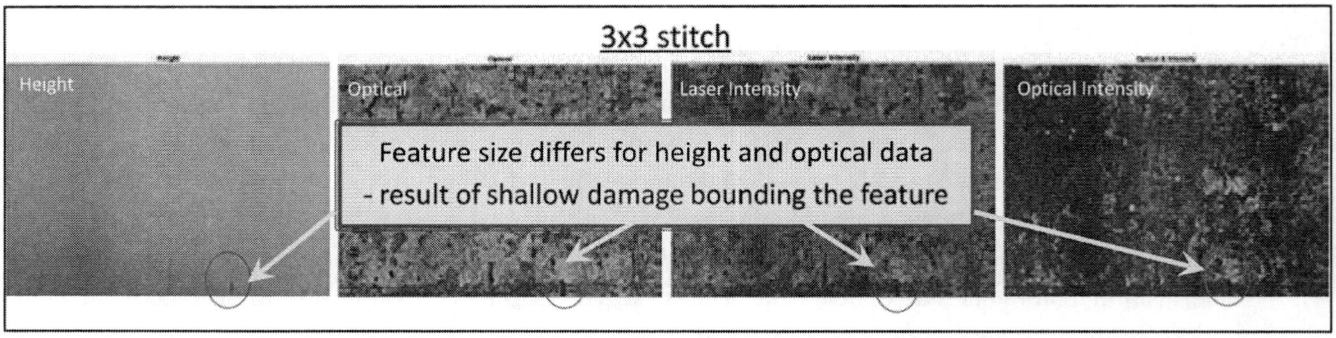

Figure 10 Difference in apparent size of a feature relative to height vs optical data. Each figure is a 3x3 mosaic.

Machine Learning Implementation

The software discussed thus far computes properties of physical features and permits user defined flagging based on specified threshold values. The next stage is the implementation and training of supervised machine learning algorithms for corrosion detection by using the output of the software along with labeled images of corrosion. The methodology will follow the path shown in Figure 11.

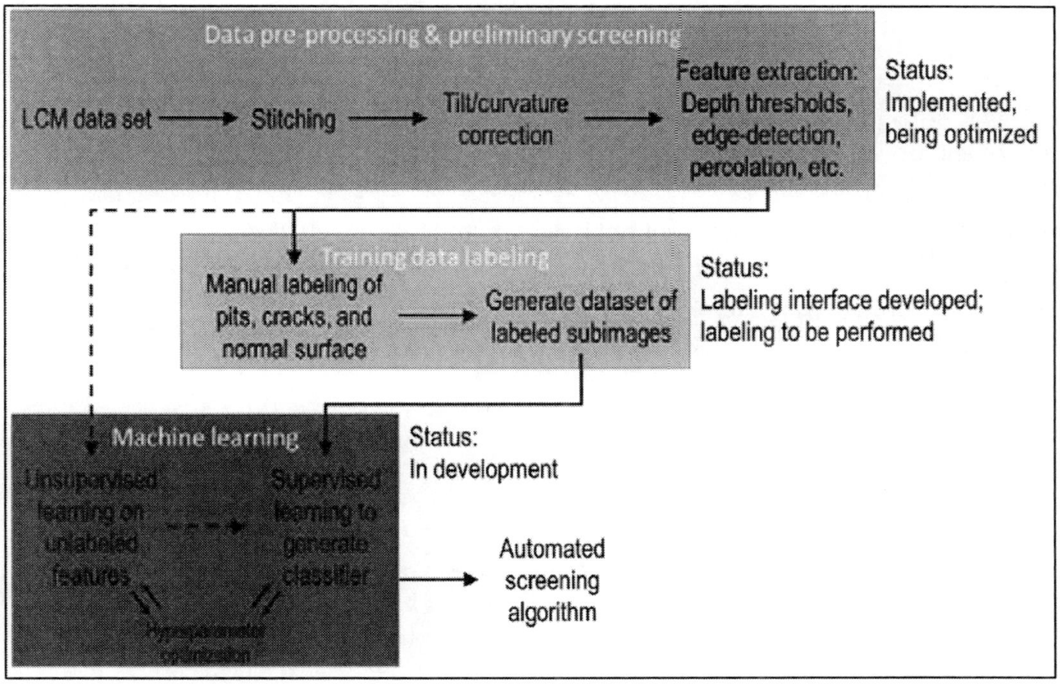

Figure 11 Methodology for application of machine learning for 3013 corrosion data.

Conclusions

This paper describes the methods used to prepare 3013 image data for application of machine learning algorithms capable of identifying corrosion events. A GUI, based on Matlab® software, was used to recover height and optical features from binary LCM output files for corrosion imaging. The software processed the sets of LCM features to compute areas, depths and shapes of surface structures, along with areal density of the structures and their associated color patterns. The structures were classified as pits, depressions, cracks and RGB functions. User specified thresholds were used to flag structures of interest, indicative of corrosion. Flagging suspect structures will significantly reduce the amount of time required for manual inspection.

As seen from Figure 10, the ability to detect and gauge surface structures is dependent on which feature data is employed. Likewise, it was found that pits could be readily identified using height data, but optical RGB data was necessary to identify cracks. Searching for cracks relied on methods used to define edges in optical images. To find the most suitable approach for identifying cracks, a number of edge detection methods were tested, including: differencing masks, filters and covariance thresholds. While covariance thresholds appeared superior in some cases, they relied on guessing the correct threshold values. If the thresholds are not properly set, either the edge would be missed, or many non-edge structures become flagged.

The software discussed in this paper is a prelude to the application of machine learning algorithms. Complex features, constructed from the LCM data, are supplied as input to the machine learning algorithms. The platforms for machine learning will be Tensorflow and Caffe. However, it is recognized that application of Tensorflow may require implementation of a Matlab-Python interface. Training sets for supervised learning will be drawn from labeled 3013 images and from coupon data.

References

1 Matlab®, Mathworks Inc.

2 Managed by Department of Nanometrology, Czech Metrology Institute. Gwyddion. [Online] [Cited: February 20, 2019.] http://gwyddion.net/.

3 Garcia, Damien. Mathworks File Exchange. [Online] [Cited: February 20, 20019.] https://www.mathworks.com/matlabcentral/fileexchange/25634-smoothn?focused=8189808&tab=function.

549059 Visualizing Chemical Process Data Via Dimensionality Reduction

The complexity of a chemical plant often requires multivariate analyses to identify the root cause of process interruptions, key variables influencing product quality, etc. Visualization of the dataset early in the data analytics journey can provide preliminary insights to guide an efficient analysis. Dimensionality reduction techniques, such as principal component analysis, provide a concise visualization in just a few dimensions. The techniques differ in how they attempt to preserve local or global properties of the entire dataset in mapping from the original, high dimensional space to the reduced, low dimensional space. The visualization readily enables cluster analysis in the reduced space that can assist in quickly identifying the key performance indicators. While there is an abundance of dimensionality reduction techniques, most are seldom applied to chemical process data. Therefore, we present case studies in which we analyze process data with a variety of these techniques, such as Isomap, t-SNE, and Laplacian Eigenmaps, among others. We outline their strengths and weaknesses and make comparisons against the industry standard approach – principal component analysis.

References:

L. van der Maaten, E. Postma, J. van den Herik. Dimensionality Reduction: A Comparative Review. *Tilburg University Technical Report*, **2009**

Z. Ge, Z. Song, S.X. Ding, B. Huang. Data Mining and Analytics in the Process Industry: The Role of Machine Learning. *Access IEEE*. **2017**.

551666 Minimizing Batch Cycle Time Using Evolutionary Design of Dynamic Experiments

Due to the nonlinear characteristics of batch processes, optimizing their performances, such as minimizing the cycle time, usually relies on a knowledge-driven model consisting of a set of algebraic differential equations. However, as the configuration and behavior of industrial process varies one from another, the knowledge-driven model is not always available. In such a case, a data-driven model which can be developed after a limited number of experiments is an attractive alternatives. Here we apply the Design of Dynamic Experiments (DoDE)[1] methodology to model the process behavior and minimize the batch cycle time using the obtained model. DoDE generalizes the classic design of experiment approach and allows time-varying variables, such as feeding profile of reactant, to be considered in the experimental design.

The DoDE approach has been applied to optimize an industrial simulation at Dow[2]. The obtained optimal operating condition was very promising but some safety and product quality constraints were violated of some designed experiments. To satisfy these constraints while running the experiments, we apply the evolutionary operation of DoDE[3] in this presentation. This methodology also reduces the initial number of experiments. The initial design is selected conservatively in the small vicinity of the previous operating conditions. After the initial data-driven model has been estimated using the collected data, an optimal operating condition with satisfying uncertainty statistics[4] is selected and implemented. The new optimal condition also serves as a new data point that will be applied to update the data-driven model. An updated optimal operating condition will then be selected to further improve the process performance. The above steps are iterated until the best process performance is achieved. We examine the evolutionary DoDE approach in an industrial simulation of polymerization process at Dow.

1. Georgakis, C., Design of Dynamic Experiments: A Data-Driven Methodology for the Optimization of Time-Varying Processes. *Industrial & Engineering Chemistry Research* **2013,** *52* (35), 12369-12382.
2. Georgakis, C.; Chin, S.; Hayot, P.; Wassick, J.; Chiang, L. H. In *Optimizing an Industrial Batch Process Using the Design of Dynamic Experiments Methodology*, AICHE Spring Meeting, Houston,TX, April 10-14; Houston,TX, 2016.
3. Wang, Z.; Georgakis, C. In *Data-Driven Optimization Using an Evolutionary Design of Dynamic Experiments for Biopharmaceutical Processes*, AICHE Annual Meeting, San Francisco, CA, Nov 13-18; San Francisco, CA, 2016.
4. Wang, Z.; Georgakis, C., An in silico evaluation of data driven optimization of biopharmaceutical processes. *AIChE Journal* **2017,** *63* (7), 2796-2805.

AUTHOR INDEX

AUTHOR INDEX

AIChE
120 Wall Street, FL 23
New York, NY 10005-4020

CURRAN ASSOCIATES INC.
proceedings.com

ISBN 978-1-5108-8681-0
9 781510 886810